Efficiency in Kinesiology: Innovative Approaches in Enhancing Motor Skills for Athletic Performance

Efficiency in Kinesiology: Innovative Approaches in Enhancing Motor Skills for Athletic Performance

Editor

Diego Minciacchi

MDPI

Basel • Beijing • Wuhan • Barcelona • Belgrade • Novi Sad • Cluj • Manchester

Editor
Diego Minciacchi
Department of Experimental
and Clinical Medicine
University of Florence
Florence
Italy

Editorial Office
MDPI
St. Alban-Anlage 66
4052 Basel, Switzerland

This is a reprint of articles from the Special Issue published online in the open access journal *Journal of Functional Morphology and Kinesiology* (ISSN 2411-5142) (available at: www.mdpi.com/journal/jfmk/special_issues/79J7HZ0C13).

For citation purposes, cite each article independently as indicated on the article page online and as indicated below:

Lastname, A.A.; Lastname, B.B. Article Title. *Journal Name* **Year**, *Volume Number*, Page Range.

ISBN 978-3-0365-8779-0 (Hbk)
ISBN 978-3-0365-8778-3 (PDF)
doi.org/10.3390/books978-3-0365-8778-3

Contents

Journal of
Functional Morphology and Kinesiology

MDPI

Editorial

Efficiency in Kinesiology: Innovative Approaches in Enhancing Motor Skills for Athletic Performance

Vincenzo Sorgente * and Diego Minciacchi

Kinesiology and Motor Control (Ki.Mo.Co.) Laboratory, Department of Experimental and Clinical Medicine, Physiological Sciences Section, University of Florence, 50134 Florence, Italy; diego.minciacchi@unifi.it
* Correspondence: vincenzo.sorgente@unifi.it

Citation: Sorgente, V.; Minciacchi, D. Efficiency in Kinesiology: Innovative Approaches in Enhancing Motor Skills for Athletic Performance. *J. Funct. Morphol. Kinesiol.* **2023**, *8*, 111. https://doi.org/10.3390/jfmk8030111

Received: 31 July 2023
Accepted: 3 August 2023
Published: 4 August 2023

The inaugural edition of the Special Issue titled "Efficiency in Kinesiology: Innovative approaches in enhancing motor skills for Athletic Performance" has been effectively concluded.

In recent years, sport science research has substantiated the efficacy of certain modalities as the paramount outline for assessing, enhancing, and even prognosticating athletic performance [1,2]. These methodologies encompassed both bio-motor (e.g., power development) and related technical components (e.g., vertical jump during a basketball shoot). However, the scientific and technological progress keeps giving rise to further potential opportunities for optimizing the core principles which craft the best sporting performances. In order to corroborate this enhancement, various approaches have been employed, spanning from the ecological, lab-independent, cost-effective, to the minimal invasiveness ones. The perspective of upgrading well-known and validated applications is indeed intriguing to the scholarly community as well as the in-field practitioners (i.e., trainers and athletes), both dedicated to the unceasing improvement of athletic conditioning [3]. Thus, the conceptual groundwork of the present volume is conceived on the need to address these novel strategies and proposals through a methodical and cohesive fashion. This Special Issue welcomes 13 original research articles plus one case report, centered on implementing cutting-edge approaches to efficiently sharpen motor skills in the function of elevating sporting performances. The athletic domains covered by this Special Issue include soccer (situational performances for both male and female competitors), swimming (exercise physiology), wrestling (sociocultural aspects), basketball (new test validation), volleyball (new test validation and exercise kinematics), handball (exercise kinematics), fencing (visual strategies), other than non-sport specific biomechanics, strength and conditioning, and robotics.

Present soccer demands are increasing in terms of running requirements and the augmenting number of scheduled matches provide several periods of fixture congestion during the season. Strategic and hyper-specific use of the team's resources is becoming a must for competitive success. With the use of GPS technology, Muñoz-Castellanos et al. [4] monitor accumulated workloads during the season, seeking for differences among roles played, and between starters versus substitutes. They find that each position (central defender, full back, midfielder, wide midfielder, striker) shows specific behaviors in distance covered during a congested competitive period. The Authors conclude that coaches should pay attention to the fatigue produced by the number of high decelerations and that an individualized training protocol should be considered according to the running requirements of each position on the pitch. Additionally, Furtado Mesa et al. [5] show that seasonal accumulated total distance, sprints, and high-speed distance are significantly greater for starters than substitutes. Also, accumulated training load and training load per minute played in matches do not differ between starters and substitutes as the accumulated training load profiles of substitutes is similar to that of starters.

Evaluating force–velocity characteristics on dry land is of the utmost importance in swimming, since higher levels of these bio-motor abilities positively affect in-water performance. Sorgente et al. [6] search for differences among swimmers' stroke (butterfly, backstroke, breaststroke, front-crawl) and distance specialization (50, 100, and 200 m) by measuring the maximum force–velocity exertion during the pull-up motion. Assessments are performed (via linear encoder) before and after taking part in an official swimming race. Both force and velocity are significant predictors of swimming race time. Sprinters (50 m and 100 m) of all strokes exert significantly higher force–velocity compared with 200 m swimmers. Interestingly, breaststroke sprinters present significantly lower force–velocity compared to sprinters specialized in the other strokes. Searching for the role of stroke and distance specializations in modeling swimmers' force–velocity abilities can heavily influence swimming training and performance.

Employing the Parental Support Scale for Children in Sports, Biletic et al. [7] investigate the effects of age and popularity of wrestling in influencing perceived parental support. During the period of entry into specialization, children perceive less parental support and lower parental belief in the benefits of sport practicing. Moreover, as could be expected, in environments in which wrestling is popular, parents know the sport better and can actively participate; therefore, children perceive more parental support. These outcomes may help coaches to better understand the athlete–parent relationships and correlated psychological aspects during sensible stages of the young athlete's development.

Jumping ability in basketball is assessed using standardized vertical jump tests which, however, lack specificity by not considering the player's basketball skills. Theodorou et al. [8] propose the pivot step jump test (PSJT) as a novel test designed to evaluate the jumping abilities of basketball players by combining a pivot step on one leg with a maximum bilateral vertical jump. To this scope, intra-and intersession reliability and validity are evaluated (performing the PSJT and a series of criterion jumping tests). No changes are found in PSJT performance between test sessions and excellent intra- and intersession reliability was observed. Furthermore, correlation coefficients indicate high factorial validity between the jumping tests and PSJT. Therefore, PSJT offers a valid assessment of jumping ability in basketball, having the practical potential to assess sport-specific jumping skills in young basketball players.

Another innovative test is proposed in this Special Issue for the sport of volleyball. Đolo et al. [9] aim to determine the test–retest reliability and discriminative ability of five sport-specific kinesthetic differentiation tests in volleyball female players. In particular, kinesthetic differentiation ability is determined by evaluating (1) overhead passing, (2) forearm passing, (3) float service with a net, (4) float service without a net, (5) float service 6 m from the net. Parameters of the intraclass correlation coefficient are excellent in all tests except for the float service with the net, whose reliability was good. Hence, the Authors endorse this specific battery test as a reliable tool to monitor kinesthetic differentiation ability in female volleyball players.

The second volleyball-centered paper of this Special Issue focuses on the relationships among ankle flexibility, knee extensors torque, and performance in countermovement jump (CMJ) by Panoutsakopoulos and Bassa [10]. Testing includes the CMJ with–without an arm swing, and—on an isokinetic dynamometer—maximal knee extensions and flexions at three angular velocities. CMJ height and relative power are positively correlated with the extensors' torque at 180°/s and are negatively correlated with the flexibility level of the dominant ankle, also revealing that more flexible players jump significantly higher during the CMJs. The Authors conclude that a more flexible ankle joint and a higher isokinetic knee extensor's torque generating capacity result in higher CMJ performance. Therefore, training of ankle flexibility should be emphasized, and specific screening should be included during preseason in youth female volleyball players.

Sport-specific kinematics is further explored in handball. In particular, Burger et al. [11] study the kinematic parameters of single side feint movement between elite and professional level handball players. In handball, the feint movement is a fundamental technical

and strategical element for offensive players to outplay their guard and score. The kinematic analysis is conducted using a GAIT—LaBACS software system (ver. 1.0), considering seven kinematic parameters for the "feint" and "actual" phases recorded by BASLER-402-FC and PANASONIC VW-D5100 cameras. Two variables have significant differences between elite and professional players: (1) step length of the stride leg; (2) moving the leg opposite to the throwing arm, demonstrating that less skilled players use more space for the same technical element.

Advancements in technology enable quantification of wide-ranging features of human movement. At variance from the study by Theodorou et al. [7] which quantifies sport-specific abilities of basketball players, here Philipp et al. [12] investigate technologies' reliability prior to comparison with established industry gold standards. The Authors seek to determine the inter-device reliability between two identical markerless motion capture systems placed in close proximity (3D-MCS, DARI Motion; each composed by eight high-definition cameras recording at 60 fps). The test comprises 29 different elementary movement tasks. The results indicate negligible or small between-device effect sizes, while showing mostly excellent, moderate, or better agreement when looking at the ICC values, and little differences as for metrics measuring joint angles and distance measures. The preliminary though promising nature of the data leads the Authors to suggest that 3D-MCS may provide practitioners with a new opportunity to measure the movement characteristics of athletes reliably and efficiently.

Innovative approaches for assessing and ameliorating sport-specific performance are inquired in fencing. Bagot et al. [13] investigate the visual activity of fencers in conditions resembling official competitions. Eight national level fencers are recruited. Measures are performed by a head-mounted Pupil Invisible Eye tracking device (Pupil Labs®, Berlin, Germany) during the simulated bouts. Findings indicate that the main fixation in foil and sabre is the upper torso, while in epee, it is the lower torso. Two additional areas of interest are identified: (1) the score machine; (2) an area involving fixations that do not target a specific area of their opponent. Although these two areas are not directed towards the opponent, they still testify to a visual activity performed during a competition and potentially reflect what happens in a real match. Conversely, the study finds no direct link between visual activity and performance. The Authors conclude that fencers adapt their visual search strategy to the fencing specialization, i.e., the ruleset, that they take part in.

Previous research investigating the association of strength performance and anthropometric variables is often performed in a sample of pooled sexes or one sex only or by utilizing tests with low ecological validity. As such, Falch et al. [14] conducted a randomized cross-over study investigating the association of anthropometrics with strength performances in the squat and bench press for resistance-trained adult males and females and whether the association differed between the sexes. Participants are tested for strength performances with 60% of their 1-RM in the squat and bench press. The Authors find that some associations between strength performance and anthropometric variables differ between males and females. Namely, in the AMRAP (as many reps as possible) squat, thigh length is inversely associated with performance in males, while fat percentage is inversely associated with performance in females. However, for both sexes, lean mass and body height are associated with 1-RM strength in the squat and bench press, while body height is inversely associated with AMRAP performance.

Apart from squat jumps, CMJ, and drop jumps, differences among other jump variations are not sufficiently known, making it difficult to select data-driven exercises. To address this gap, Janikov et al. [15] compare specific concentric and eccentric jump parameters of maximal effort CMJ, jumps over 50 cm hurdle (HJ), and jumps onto a 50 cm box (BJ). The data (average of the three repetitions of each jump, performed on separate days) are collected using force platforms and a linear position transducer. No differences are found in peak velocity, peak vertical and resultant force, and total impulsion time. The Authors stress how overall training load could decrease dramatically when performing BJ, because of the half-reduced peak impact force compared to CMJ and HJ.

Schönau et al. [16] investigate whether the amplitude-force relationship of back muscles could be altered systematically by using different training modalities. Graded submaximal forces on the back are applied by defined forward tilts in a full-body training device (CTT Centaur, BfMC). Surface EMG is recorded utilizing a monopolar 4 × 4 quadratic electrode scheme in the lower back area. Between-group tests reveal significant differences between strength trained subjects vs. both endurance-trained and physically inactive subjects, at both the medial and caudal electrode positions. These results point towards training-related changes to the fiber-type composition of muscles in the strength-trained participants, particularly for their paravertebral region.

The Special Issue closes with an insight into the innovative use of robotics for neuromotor rehabilitation in clinical settings. Koseki et al. [17] provide an ankle rehabilitation training program with robot-assisted device in a patient with incomplete spinal cord injury. Using a three-dimensional motion analyzer and surface EMG, the Authors evaluate the treatment effectiveness using ankle plantar dorsiflexion exercises in the sitting position, knee flexion—extension exercises in the standing position, and stepping exercises in the standing position with hybrid-assistive limb assistance. Before rehabilitation, the patient was unable to perform voluntary ankle movements due to severe motor–sensory dysfunction. The training program is able to induce muscle potentials in the left tibialis anterior muscle during plantar dorsiflexion of the ankle.

Finally, we wish to gratefully acknowledge the essential contributions from all the Authors, Reviewers, and Editors towards this Special Issue.

Given the great success of this Special Issue, we have launched a second edition. We believe that this subject holds the potential to drive advancements in sports science by bridging cutting-edge scientific research with the on-field training methodologies and experiences.

Conflicts of Interest: The authors declare no conflict of interest.

References

1. Quarta, E.; Cohen, E.J.; Bravi, R.; Minciacchi, D. Future Portrait of the Athletic Brain: Mechanistic Understanding of Human Sport Performance Via Animal Neurophysiology of Motor Behavior. *Front. Syst. Neurosci.* **2020**, *14*, 596200. [CrossRef] [PubMed]
2. Sperlich, B.; Aminian, K.; Düking, P.; Holmberg, H.-C. Editorial: Wearable Sensor Technology for Monitoring Training Load and Health in the Athletic Population. *Front. Physiol.* **2020**, *10*, 1520. [CrossRef] [PubMed]
3. Muñoz de la Cruz, V.; Agudo-Ortega, A.; Sorgente, V.; Turner, A.P.; González-Ravé, J.M. The Effectiveness of Adjusting Resistance Training Loads through Velocity-Based Techniques in Experienced Sprinters: A Case Series Study. *Front. Physiol.* **2023**, *14*, 1241459. [CrossRef]
4. Muñoz-Castellanos, B.; Rabano-Muñoz, A.; Requena, B.; Suarez-Arrones, L.; Asian-Clemente, J.A. Congested Period in Professional Youth Soccer Players Showed a Different High Decelerations Profile in the Group Performance and a Specific Positional Behaviour. *J. Funct. Morphol. Kinesiol.* **2022**, *7*, 108. [CrossRef] [PubMed]
5. Furtado Mesa, M.; Stout, J.R.; Redd, M.J.; Fukuda, D.H. Accumulated Workload Differences in Collegiate Women's Soccer: Starters versus Substitutes. *J. Funct. Morphol. Kinesiol.* **2023**, *8*, 78. [CrossRef] [PubMed]
6. Sorgente, V.; Agudo-Ortega, A.; Lopez-Hernandez, A.; Santos del Cerro, J.; Minciacchi, D.; González Ravé, J.M. Relationship between Maximum Force–Velocity Exertion and Swimming Performances among Four Strokes over Medium and Short Distances: The Stronger on Dry Land, the Faster in Water? *J. Funct. Morphol. Kinesiol.* **2023**, *8*, 20. [CrossRef] [PubMed]
7. Biletic, I.; Karnincic, H.; Baic, M. Effects of Age and Popularity of Sport on Differences among Wrestlers' Parental Support: An Exploratory Study. *J. Funct. Morphol. Kinesiol.* **2023**, *8*, 65. [CrossRef]
8. Theodorou, A.S.; Rizou, H.P.; Zacharakis, E.; Ktistakis, I.; Bekris, E.; Panoutsakopoulos, V.; Strouzas, P.; Bourdas, D.I.; Kostopoulos, N. Pivot Step Jump: A New Test for Evaluating Jumping Ability in Young Basketball Players. *J. Funct. Morphol. Kinesiol.* **2022**, *7*, 116. [CrossRef] [PubMed]
9. Đolo, K.; Grgantov, Z.; Kuvačić, G. New Specific Kinesthetic Differentiation Tests for Female Volleyball Players: Reliability, Discriminative Ability, and Usefulness. *J. Funct. Morphol. Kinesiol.* **2023**, *8*, 63. [CrossRef] [PubMed]
10. Panoutsakopoulos, V.; Bassa, E. Countermovement Jump Performance Is Related to Ankle Flexibility and Knee Extensors Torque in Female Adolescent Volleyball Athletes. *J. Funct. Morphol. Kinesiol.* **2023**, *8*, 76. [CrossRef] [PubMed]
11. Burger, A.; Vrdoljak, D.; Foretić, N.; Spasić, M.; Pavlinović, V. Differences between Elite and Professional Male Handball Players in Kinematic Parameters of Single Fake Movement. *J. Funct. Morphol. Kinesiol.* **2023**, *8*, 47. [CrossRef] [PubMed]

12. Philipp, N.M.; Cabarkapa, D.; Cabarkapa, D.V.; Eserhaut, D.A.; Fry, A.C. Inter-Device Reliability of a Three-Dimensional Markerless Motion Capture System Quantifying Elementary Movement Patterns in Humans. *J. Funct. Morphol. Kinesiol.* **2023**, *8*, 69. [CrossRef] [PubMed]
13. Bagot, P.; Fournier, J.F.; Kerivel, T.; Bossard, C.; Kermarrec, G.; Martinent, G.; Bernier, M. Visual Search Strategies of Elite Fencers: An Exploratory Study in Ecological Competitive Situation. *J. Funct. Morphol. Kinesiol.* **2023**, *8*, 106. [CrossRef]
14. Falch, H.N.; Haugen, M.E.; Larsen, S.; van den Tillaar, R. Association of Strength Performance in Bench Press and Squat with Anthropometric Variables between Resistance-Trained Males and Females. *J. Funct. Morphol. Kinesiol.* **2023**, *8*, 19. [CrossRef] [PubMed]
15. Janikov, M.T.; Pádecký, J.; Doguet, V.; Tufano, J.J. Countermovement, Hurdle, and Box Jumps: Data-Driven Exercise Selection. *J. Funct. Morphol. Kinesiol.* **2023**, *8*, 61. [CrossRef] [PubMed]
16. Schönau, T.; Anders, C. EMG Amplitude–Force Relationship of Lumbar Back Muscles during Isometric Submaximal Tasks in Healthy Inactive, Endurance and Strength-Trained Subjects. *J. Funct. Morphol. Kinesiol.* **2023**, *8*, 29. [CrossRef] [PubMed]
17. Koseki, K.; Takahashi, K.; Yamamoto, S.; Yoshikawa, K.; Abe, A.; Mutsuzaki, H. Use of Robot-Assisted Ankle Training in a Patient with an Incomplete Spinal Cord Injury: A Case Report. *J. Funct. Morphol. Kinesiol.* **2023**, *8*, 31. [CrossRef] [PubMed]

Journal of
Functional Morphology and Kinesiology

MDPI

Article

Visual Search Strategies of Elite Fencers: An Exploratory Study in Ecological Competitive Situation

Pierre Bagot [1], Jean F. Fournier [2], Thibault Kerivel [1], Cyril Bossard [1], Gilles Kermarrec [1], Guillaume Martinent [3,*] and Marjorie Bernier [1]

1 Centre de Recherche sur l'Éducation, l'Apprentissage et la Didactique, University Brest, F-29200 Brest, France; pierre.bagot@univ-brest.fr (P.B.)
2 Laboratoire Interdisciplinaire en Neurosciences, Physiologie et Psychologie: Activité Physique, Santé et Apprentissages, University Paris Nanterre, F-92001 Nanterre, France
3 Laboratoire sur les Vulnérabilités et l'Innovation dans le Sport, University Lyon 1, F-69622 Lyon, France
* Correspondence: guillaume.martinent@univ-lyon1.fr

Abstract: This study investigates the visual activity of fencers in conditions resembling official competitions. Previous research in experimental conditions has shown that experts focus on specific areas of the torso and the armed arm to control movement initiation. Eight right-handed fencers (epee: two males, one female; foil: one male; sabre: two males, two females) participated in a simulated competition, wearing an eye tracker during one bout. The findings showed that the main fixation in foil and sabre is the upper torso, while in epee, it is the lower torso. In epee and sabre, the upper torso is viewed about 50% of the time, with three other areas also observed, while in foil, the fixation is totally directed to the upper torso. Additionally, two new areas of interest were identified: the score machine and an area involving fixations other than the opponent. The study found no direct link between visual activity and performance. The visual search strategy varies among weapons, with foil using a gaze anchor or foveal spot and epee and sabre utilizing a visual pivot due to the discipline's inherent rules. The study also emphasizes that competition-like conditions can disrupt visual activity with external stimuli, possibly affecting performance.

Keywords: fencing; visual search strategies; competitive situation; eye tracking; elite

Citation: Bagot, P.; Fournier, J.F.; Kerivel, T.; Bossard, C.; Kermarrec, G.; Martinent, G.; Bernier, M. Visual Search Strategies of Elite Fencers: An Exploratory Study in Ecological Competitive Situation. *J. Funct. Morphol. Kinesiol.* **2023**, *8*, 106. https://doi.org/10.3390/jfmk8030106

Academic Editor: Diego Minciacchi

Received: 9 May 2023
Revised: 24 July 2023
Accepted: 25 July 2023
Published: 27 July 2023

1. Introduction

Perceptual-cognitive skills play a crucial role in competitive sport [1], particularly in combat sports (e.g., karate, boxing, fencing), where athletes are constantly anticipating the forthcoming attack of their opponent based on prior cues or information available from the opponent's behavior [2]. Athletes in combat sports fight at close range and require superior perceptual abilities to adapt to opponents' attacks [3,4]. In order to anticipate opponents' attacks, react, and respond with quickness and accuracy, fighters must perceive valuable information from a large quantity of dynamically shifting information about the competition. This information may be derived from a variety of sources, such as the opponent's movements and posture, the distance to the opponent, or the match's status [2]. Such information could be recognized and processed by athletes through visual activity, allowing them to foresee the behavior of their opponents and make decisions more suitable to winning the game.

In fencing, an average action to score a point lasts 5 s in foil and 15 s in epee [5], so a quick and adequate reaction to an opponent's actions is one of the main determinants of performance in fencing. Therefore, for fencers, the method of perceiving information from the environment (i.e., kinematic information from the opponent) plays a crucial part in the effectiveness of the technical and tactical actions. In sport domains, visual activity has been investigated widely over the past two decades. Several meta-analyses [6,7] have indicated that the visual activity of experts differs when compared to novices. Particularly, experts have fewer and longer fixations than novices. The presence of systematic differences in gaze behavior between experts and novices is consistent in studies in different sports. Few studies have focused on the visual search strategies (i.e., combination of different variables: area of interest (AOI), fixation duration, number of fixations, saccades duration) of fencers. One of the first studies came from Hagemann et al. [8] who examined the eye movement of 15 expert fencers, 15 advanced fencers, and 32 sports students. In laboratory settings, using the spatial occlusion paradigm, participants were invited to sit and watch fencing attacks on a computer screen (first-person point of view). After this, the occlusion participants had to click on the anticipated target area. Three eye-tracking variables were recorded: viewing time (dwell time %), fixation duration, and number of fixations in each video clip and on each AOI. The results showed that fencers of all performance levels fixated predominantly on the trunk and on the opponent's weapon, but expert fencers recorded longer fixations than advanced fencers and sport students on the upper trunk region. Also, novice fencers tended to fixate much longer on the upper legs of their opponent compared to advanced and elite fencers. When the trunk of the opponent was occluded from the clips, all participants changed their visual activity from the trunk to adjacent areas. Advanced fencers and sport students recorded a significant decrease in performance (% prediction), while expert fencers' performance prediction did not change, demonstrating an expertise-level effect. Another study, where an expert vs. novice paradigm was used, indicated that experts use different visual perception strategies than novices [9]. Nineteen participants separated into two groups (experts vs. beginners) took part in this study. Each participant fenced two opponents, one left-handed and one right-handed. The results showed that novices tend to fixate much longer on the opponent's weapon compared to expert fencers. Moreover, novices spent an equivalent amount of time looking at five different areas (guard, foil, armed hand, lower trunk, and upper trunk), while experts spent significantly more time on the upper trunk and the armed hand.

Witkowski et al. [10,11] conducted a series of studies on the impact of opponent's handedness on visual search strategies of fencers, with the hypothesis that experts use different visual search strategies depending on the opponent's handedness. In the first study [10], 12 expert foil fencers were invited to fight two opponents, one with right-handedness and the other with left-handedness, during 20 s duels. The results showed that when facing left-handed opponents, experts tended to fixate more often and much longer on the armed hand of their opponents than the other areas (guard, foil, armed hand, lower trunk, and upper trunk). Additionally, when facing a left-handed opponent, experts spent an equal amount of time staring at the armed hand and the upper torso. When facing a right-handed opponent, experts fixated more often and spent more time on the upper torso than when facing left-handed opponents. In addition, when facing a right-handed opponent, experts fixated more often and spent more time fixating on the upper torso than any other body area. Another study by Witkowski et al. [11], also using the opponent's handedness paradigm, was conducted to find out if attacking and defensive actions had an impact on visual search behavior. Twelve female foil experts took part in this research. For each participant, the study involved two tasks, attack and defense, and two conditions, a right- and a left-handed opponent. Each participant performed 10 repetitions of each task under each condition, which altogether amounted to 40 actions. The results showed that during offensive actions, foil fencers spent more time looking at the armed hand and generated a higher number of fixations to this armed hand when facing a left-handed opponent (compared to a right-handed opponent). Moreover, in fights versus left-handed opponents, the armed hand attracted the most fixations compared to other areas of interest (AOIs). The same result was found for defensive actions. During bouts with left-handed opponents, foil fencers spent more time observing and made more fixation on the armed hand. Facing a right-handed competitor makes foil fencers change their visual search strategies. The results showed that the upper torso attracts a higher number of fixations in attack and more fixations and longer observation times on defense than when facing left-handed opponents. Those results were explained by the fact that facing left-handed competitors is less frequent and, thus, they are viewed as less predictable in their actions. These results could be explained by an increase in anxiety that may influence the stimulus-driven attentional system (bottom-up) over the goal-oriented attentional system (top-down) (corresponding with the Attentional Control Theory [12]) and, consequently, may boost the level of attention directed toward threat-related stimuli.

Taken together, these studies help to elucidate the visual search strategies used by expert fencers. Two methodologies were used to investigate these phenomena. The first included an experimental methodology [8], where fencers were sitting in front of a screen and using a joystick to respond. The other one, with more ecological conditions, based on research conducted by Witkowski and colleagues [9–11], collected fencer's visual activity directly during a fight. Nevertheless, the ecological nature of this research should be questioned. Two of the three previously cited studies found limitations in the choice of action to be performed [10] or the duration of the bout [11]. In research conducted in 2018 [10], with the fight duration limited to 20 s, it is possible that the fencer's activity was influenced by being required to perform an action within a predetermined time frame. In addition, there is no indication of how points are calculated. Is was a legitimate duel that ended when the first fencer reached 15 points, or was there no scoring recorded? This issue has a significant impact on performance due to (i) the intensity with which the fencer engages in the combat and (ii) the cumulative effect of stress on performance. Even though a 2020 study [11] was conducted on a piste, the actions requested were forced by the protocol itself (i.e., 10 offensive actions and then 10 defensive actions against a left-handed opponent and then a right-handed opponent, with a balance between the two conditions). This type of protocol, despite being ecological as it collects data directly from participants in action but in a controlled situation, is different from a real competition. Indeed, in a real bout, (i) the duration of a point can range from under a second (an action is performed immediately after the "aller") to 60 s or more [5], and (ii) attacking, defensive, or counter-attacking actions are not predetermined and are more likely to be produced in the stream of the duel with power relations at play between opponents.

The present study aimed to expand our understanding of the visual activity of fencers by proposing an ecological protocol, quite similar to what expert fencers experienced during competition. To this end, we intended to examine the visual activity of fencers during a simulated competition. To date, no study has investigated the visual activity of fencers during a simulated competition. Therefore, the aim of the present study was to examine the visual activity of fencers in situ. Specifically, with the support of the aforementioned research, we intended to investigate the possibility of various visual search strategies between weapons and between won and lost points.

2. Materials and Methods

2.1. Participants

A group of 8 right-handed fencers (epee: 2 males, 1 female; foil: 1 male; sabre: 2 males, 2 females) aged from 20 to 31 years (M = 25.88; SD = 3.87), from the French national team, volunteered to take part in this research. According to the classification of McKay et al. [13], 6 participants can be categorized as world-class athletes (Tier 5) with at least one medal at a major global championship in the last Olympic cycle (2020–2024). Additionally, those participants were ranked between the 5th and 172th place [14]. The last two participants can be categorized as elite/international athletes (Tier 4) with at least two participations at a major global championship in the last Olympic cycle (2020–2024). They were ranked between the 130th and 240th place [14]. The study's research protocol was carried out in accordance with the international ethical guidelines and data protection conditions. The study was approved by the ethics committee of Nantes University with ID number: 08042021 (8 April 2021). All participants were informed about the procedures of the study and signed the informed consent.

2.2. Materials and Measures

All measures were performed with a head-mounted Pupil Invisible Eye tracking device (Pupil Labs®, Berlin, Germany), with a sampling of 30 Hz and a recording resolution of 1088 × 1080 pixels. Recording was performed with a OnePlus 8 smartphone (OnePlus®, Shenzhen, China) worn in a waist bag and connected to the eye tracker. This system allowed for data collection in an ecological setting during a simulated competition. Pupil Invisible eye tracker was chosen because it could be worn under a fencing mask and did not require calibration [15]. Pupil Player app was used to manage and export the data. This software extracts scene video and visual activity recordings and combines them to create a video consisting of scene video and a cursor, indicating foveal vision activity. Frame-by-frame analysis was performed using Adobe Premiere Pro 2023 (Adobe®, San José, CA, USA). Each fixation was defined as the condition in which the eye remained stationary for 100 ms or three frames with a variation tolerance of approximately 1.5 degrees [4,6]. The participants' visual fields were divided into specific AOI, as outlined and analyzed by Witkowski et al. [11]. The first author carried out an analysis on 10% of the dataset before proposing it to two other researchers familiar with this type of analysis. Disagreements about AOI or delimitation of a fixation duration were discussed with regard to the theoretical ground until a consensus was reached between the three researchers. After validation of the encoding, the first author carried out his analysis, independently, on the entire dataset.

The analysis was carried out using three eye-tracking variables:

1. Fixation duration—the average length of fixation on a given area per point;
2. Fixation count—the mean number of individual eye fixations on a given area per point;
3. Dwell time—the time devoted to a given area per point expressed in percentage points.

2.3. Procedure

The study was conducted during a simulated competition that replicated an Olympic competition, specifically focusing on the second day of competition (direct elimination table). To recreate the second day of the Olympic competition, each participant engaged in five 15-point matches within a single day, with a 60 min recovery period between matches (equivalent to T64 to semi-final). The opponents were members of the French Team, ensuring that all matches replicated what fencers experience during real international competitions. Similar to actual tournaments, an official referee informed the fencers about the start and end of each point and enforced the corresponding weapon's rules. One simulated competition was organized for each weapon and gender: female epee, male epee, male foil, female sabre, and male sabre, with the exception of female foil. Due to the organization of the simulated competition, as well as the setup of the eye tracker and the discomfort experienced while wearing it, eye tracker data collection was conducted only during match 1 and match 3. Each participant was briefed on the procedure ahead of the start of the competition. After setting up the equipment, a three-point calibration was performed to ensure that the auto-calibration remained accurate with the mask on, as the eye tracker could have moved. Then, participants engaged in combat with an opponent on a piste in a well-lit fencing hall. The winner of a fencing match is the first fencer to accumulate 15 points. A point begins when the referee says "Allez" (the French word for "Go") and ends when one of the fencers scores a point that the referee validates. Following each point, both fencers must return to the center of the piste to engage in the next point. In épée and foil, the match is divided into three three-minute periods; if neither fencer reaches 15 points at the end of the three periods, the fencer with the higher score is declared the winner. In sabre, a halftime break is introduced when one fencer reaches 8 points, and there is no time limit; the winner is the first fencer to reach 15 points.

To avoid moisture from the athlete's sweat infiltrating the eye tracker, to limit the inconvenience of wearing glasses under a mask, and due to the variability in match duration, recording (points and pauses) lasted between 6 min and 14 s and 30 min and 31 s (M = 16.56 min; SD = 08.54 min).

2.4. Data Analysis

The Pupil Invisible (Pupil Labs®) Eye tracking device has a constant error margin of 4.5° [15], which represents a deviation of approximately 8 cm when the object fixated is at 100 cm and ~23.6 cm when the object fixated is at 300 cm. Additionally, due to the ecological nature of our research, the distance between fencers constantly varies during the match, so the data must be analyzed with great caution. To do so, the authors developed four figures representing the opponent's fencer at multiple distances (100 cm and 300 cm) in order to depict the AOIs with the maximum span (see Figure 1). During the analysis, the author compared, when needed, the footage with the appropriate figure to ensure the coding of the right AOI. In addition, gaze motion can help determine where the fencers are fixated. For example, for the blade, only tracking gaze motion can determine whether it is a fixation on the blade or an AOI situated behind the blade, such as the armed hand or torso. Practically, if a fixation was made on the blade and a smooth pursuit followed afterward, we considered it as a fixation on the blade, which extended from the beginning of the cursor stabilization on the blade until the start of the smooth pursuit. Conversely, if the blade moved but the cursor stayed in the same place, fixation was noted to the corresponding AOI. Smooth pursuits were not taken into consideration during the data analysis.

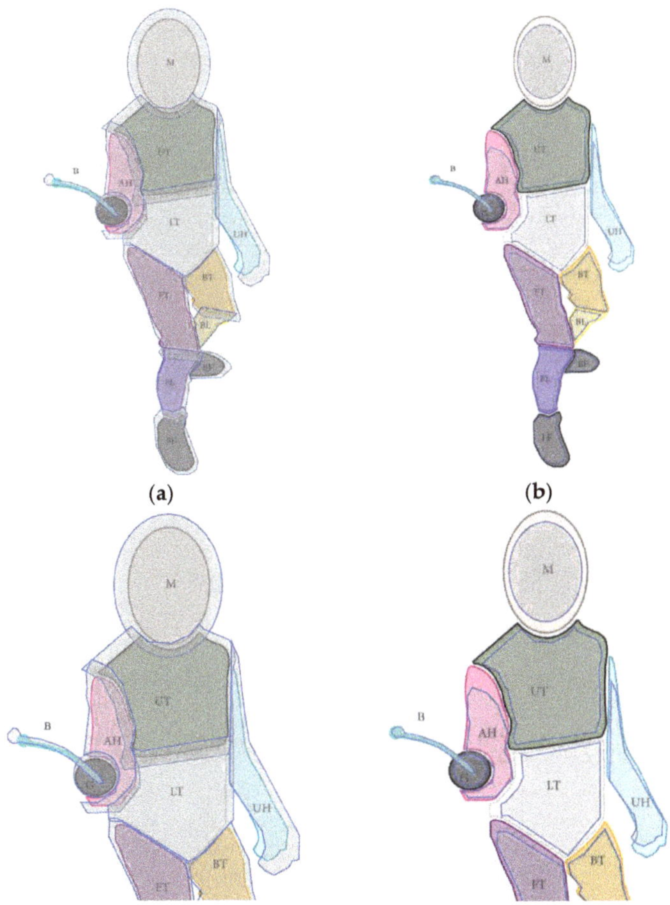

(a) (b)

(c) (d)

Figure 1. Reference image used during the data analysis. Areas of interest (AOIs) taken from Witkowski et al. [11] and maximum span (represented in red). (**a**) AOI with maximum span for a distance of 300 cm; (**b**) AOI with minimum span for a distance of 300 cm; (**c**) AOI with maximum span for a distance of 100 cm; (**d**) AOI with minimum span for a distance of 100 cm. AH: armed hand; B: blade; BF: back foot; BL: back leg; BT: back tight; FF: front foot; FL: front leg; FT: front tight; G: guard; LT: lower torso; M: mask; UT: upper torso.

The eye-tracking variables were analyzed separately. Due to the small number of participants to weapons, only descriptive statistics were processed with mean, standard deviation, minimum, and maximum. All the collected data allowed us to extract 267 points, which corresponds to an average of 26 points per participant (min = 15; max = 70). The presented results include fixation duration, number of fixations, and dwell time per point. In addition, to determine if an AOI can be considered as such, a selection criterion of 5 fixations per fencer during the whole experiment was applied.

3. Results

The results are presented in four sections: (i) the identification of "Areas Of Interest", (ii) the AOIs per point (all three weapons combined), (iii) the AOI differences between weapons, and (iv) the comparisons between won and lost point (all three weapons combined). The means and SD displayed below are those for one point. We believe that it is more interesting for professionals, trainers, and athletes to report visual activity on a single point.

3.1. *Area of Interest*

The analysis revealed 9 AOIs already highlighted by Witkowski: armed hand, blade, front foot, front leg, front thigh, guard, lower torso, mask, and upper torso. Two new AOIs were additionally identified: score machine and out of bound; they were not directly related to the opponent (see Figure 2). The first, score machine (SM), was a device that displayed the match score as well as the time remaining in the period. Moreover, a light appears whenever a fencer touches their opponent. The out-of-bound (OB) area refers to different fixations made away from the opponents, in particular a luminous device located at the end of the piste and at a height, which displays a color (green or red) as soon as a fencer touches their opponent.

3.2. *AOI Per Point*

3.2.1. Fixation Duration

For fixation duration (Table 1), we noted that, on average, the armed hand (mAH = 1195 ms; SD = 1166 ms), the lower torso (mLT = 2410 ms; SD = 3466 ms), the mask (mM = 1204 ms; SD = 1099 ms), and the upper torso (mUT = 2013 ms; SD = 1718 ms) were the AOIs where fencers looked for the longest duration during a point. Moreover, when emphasizing on maximum, we can observe that the guard (MaxG = 11,094 ms), the lower torso (MaxLT = 23,451 ms), and the upper torso (MaxUT = 11,759 ms) were the AOIs fixated more than 11 s to 23 s in a point.

Table 1. Mean, standard deviation, minimum, and maximum fixation time to particular areas of interest, in a point, expressed in ms.

AOI	Mean	SD	Min	Max
AH	1195	1166	200	7069
B	443	204	234	985
FF	412	148	217	738
FL	268	17	229	316
FT	366	211	134	1005
G	902	1208	167	11,094

Table 1. *Cont.*

AOI	Mean	SD	Min	Max
LT	2410	3466	158	23,451
M	1204	1099	211	7649
OB	313	209	134	871
SM	323	212	100	768
UT	2013	1718	141	11,759

Note. AH = armed hand; B = blade; FF = front foot; FL = front leg; FT = front thigh; G = guard; LT = lower torso; M = mask; OB = out of bounds; SM = score machine; UT = upper torso.

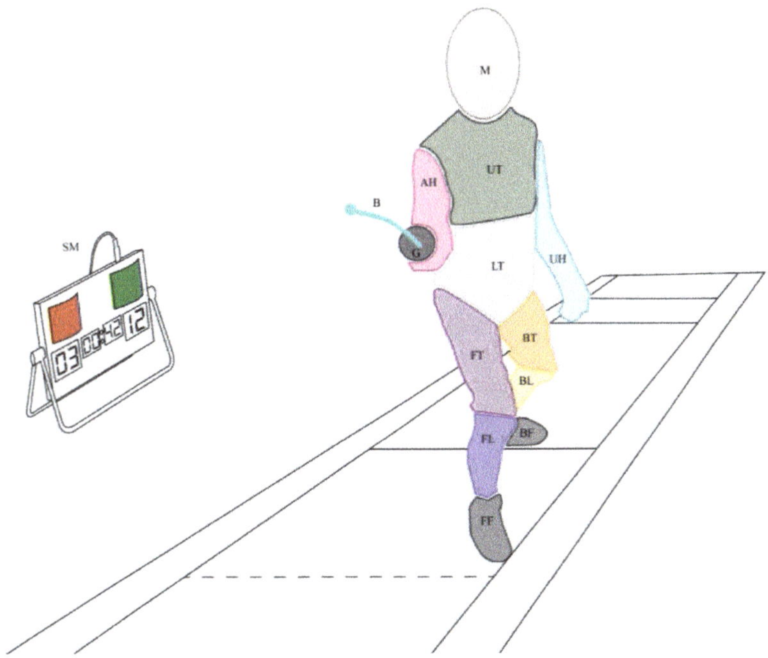

Figure 2. Reference image with areas of interest (AOIs) taken from Witkowski et al. [11] and added AOI: score machine. AH: armed hand; B: blade; BF: back foot; BL: back leg; BT: back tight; FF: front foot; FL: front leg; FT: front tight; G: guard; LT: lower torso; M: mask; SM: score machine; UT: upper torso.

3.2.2. Numbers of Fixation

For number of fixations (Table 2), we noted that the armed hand (mAH = 0.75), guard (mG = 1.34), lower torso (mLT = 1.73), and upper torso (mUT = 1.15) were the most frequent AOIs fixated by fencers during a point. All other AOIs, except mask (0.2), were fixated less than 0.1 times, on average, by fencers.

Table 2. Mean, standard deviation, minimum, and maximum number of fixations to particular areas of interest in a point.

AOI	Mean	SD	Min	Max
AH	0.75	2.26	0	16
B	0.03	0.2	0	2
FF	0.04	0.27	0	2

Table 2. *Cont.*

AOI	Mean	SD	Min	Max
FL	0.02	0.17	0	2
FT	0.04	0.23	0	2
G	1.34	4.12	0	30
LT	1.73	4.51	0	24
M	0.2	0.5	0	4
OB	0.06	0.29	0	2
SM	0.03	0.22	0	2
UT	1.15	1.57	0	13

Note. AH = armed hand; B = blade; FF = front foot; FL = front leg; FT = front thigh; G = guard; LT = lower torso; M = mask; OB = out of bounds; SM = score machine; UT = upper torso.

3.2.3. Dwell Time

For dwell time (Table 3), the upper torso (mUT = 53.75%) was the AOI where fencers devoted the most time during a point. In other words, during a point, fencers spent half of the time looking at the upper torso of their opponent. The other half was partially distributed between four AIOs: armed hand (mAH = 9.35%), guard (mG = 7.40%), lower torso (mLT = 18.10%), and mask (mM = 10.25%).

Table 3. Mean, standard deviation, minimum, and maximum time devoted to particular area of interest, in a point, expressed in %.

AOI	Mean	SD	Min	Max
AH	9.35	24.4	0	100
B	0.05	0.5	0	6.5
FF	0.10	0.5	0	4.2
FL	0.00	0.15	0	2
FT	0.45	3.85	0	65.7
G	7.40	18.55	0	98.9
LT	18.10	32.5	0	100
M	10.25	28.35	0	100
OB	0.10	0.8	0	9.7
SM	0.10	0.65	0	9.9
UT	53.75	45.3	0	100

Note. AH = armed hand; B = blade; FF = front foot; FL = front leg; FT = front thigh; G = guard; LT = lower torso; M = mask; OB = out of bounds; SM = score machine; UT = upper torso.

3.3. *Comparisons between Weapons*

3.3.1. Fixation Time

First, we noted that blade, front foot, and front leg were only fixated in epee. In foil, the front tight and score machine were not fixated. Regarding fixation duration (Table 4), in epee, the longest fixation was on the lower torso (mLT = 2542 ms; SD = 2916 ms). In foil, the longest fixated AOI was the upper torso (mUT = 2453 ms; SD = 1373 ms). In sabre, the upper torso (mUT = 2167 ms; SD = 1964 ms) and the lower torso (mLT = 2174 ms; SD = 5023 ms) were the longest AOIs fixated.

Table 4. Mean, standard deviation, minimum, and maximum fixation time to particular areas of interest by weapons, in a point, expressed in ms.

AOI	Epee				Foil				Sabre			
	Mean	SD	Min	Max	Mean	SD	Min	Max	Mean	SD	Min	Max
AH	817	785	200	3622	302	47	268	335	1584	1362	201	7069
B	493	258	234	985	/	/	/	/	/	/	/	/
FF	425	182	217	738	/	/	/	/	/	/	/	/
FL	261	37	229	316	/	/	/	/	/	/	/	/
FT	267	120	134	457	/	/	/	/	888	166	771	1005
G	1117	1756	167	11,094	390	145	201	536	539	396	168	1742
LT	2542	2916	158	16,132	863	529	503	1642	2174	5023	235	23,451
M	344	230	211	609	1139	95	1072	1206	1281	1255	235	7649
OB	245	61	200	334	687	261	503	871	214	56	134	268
SM	363	239	100	768	/	/	/	/	167	NA	167	167
UT	914	649	141	3340	2455	1373	369	5829	2167	1964	267	11,759

Note. AH = armed hand; B = blade; FF = front foot; FL = front leg; FT = front thigh; G = guard; LT = lower torso; M = mask; OB = out of bounds; SM = score machine; UT = upper torso; NA = Not Applicable.

3.3.2. Number of Fixations

In epee, as seen in Table 5, the most frequently fixated AOI was the lower torso (mLT = 6.97; SD = 7.10), but the guard (mG = 5.21; SD = 7.2) was also fixated on a substantial number of times. In foil, the most frequently fixated area, during a point, was the upper torso (mUT = 1.43; SD = 0.73). It should be noted that other AOIs in foil were fixated on, on average, less than 0.2 times per point. Finally, in sabre, the AOI that was most frequently fixated on, on average, was the upper torso (mUT = 0.84; SD = 0.72).

Table 5. Mean, standard deviation, minimum, and maximum number of fixations to particular areas of interest by weapons in a point.

AOI	Epee				Foil				Sabre			
	Mean	SD	Min	Max	Mean	SD	Min	Max	Mean	SD	Min	Max
AH	2.60	4.07	0	16	0.06	0.31	0	2	0.21	0.52	0	3
B	0.14	0.44	0	2	/	/	/	/	/	/	/	/
FF	0.18	0.53	0	2	/	/	/	/	/	/	/	/
FL	0.10	0.35	0	2	/	/	/	/	/	/	/	/
FT	0.14	0.44	0	2	/	/	/	/	0.01	0.11	0	1
G	5.21	7.20	0	30	0.16	0.54	0	3	0.13	0.36	0	2
LT	6.97	7.10	0	24	0.08	0.27	0	1	0.13	0.34	0	1
M	0.08	0.41	0	3	0.04	0.20	0	1	0.30	0.60	0	4
OB	0.08	0.33	0	2	0.06	0.31	0	2	0.05	0.27	0	2
SM	0.13	0.42	0	2	/	/	/	/	0.01	0.08	0	1
UT	1.67	2.86	0	13	1.43	0.73	1	4	0.84	0.72	0	4

Note. AH = armed hand; B = blade; FF = front foot; FL = front leg; FT = front thigh; G = guard; LT = lower torso; M = mask; OB = out of bounds; SM = score machine; UT = upper torso.

3.3.3. Dwell Time (%)

Focusing on dwell time (Table 6), we can observe that in epee, during a point, half of the time spent by the fencer fixating on an area was on the lower torso (mLT = 55.8%;

SD = 33.8%), followed by three areas: guard (mG = 23.7%; SD = 29.3%), upper torso (mUT = 10.4%; SD = 20.2%), and armed hand (mAH = 8.4%; SD = 15.2%). In foil, the upper torso was where the fencer spent most of his time during a point (mUT = 96.6%; SD = 8.3%). Lastly, in sabre, fencers spent half of their time observing the upper torso (mUT = 57.4%; SD = 44.6%). During the other half, they fixated on different AOIs: mask (mM = 17.6%; SD = 35.6%), armed hand (mAH = 12.6%; SD = 30.1%), lower torso (mLT = 8.6%; SD = 25%), and guard (mG = 2.8%; SD = 10.5%).

Table 6. Mean, standard deviation, minimum, and maximum time devoted to particular area of interest, by weapons, in a point, expressed in %.

AOI	Epee				Foil				Sabre			
	Mean	SD	Min	Max	Mean	SD	Min	Max	Mean	SD	Min	Max
AH	8.4	15.2	0	63.8	0.6	2.4	0	16	12.6	30.1	0	100
B	0.3	1.1	0	6.5	/	/	/	/	/	/	/	/
FF	0.3	1	0	4.2	/	/	/	/	/	/	/	/
FL	0.2	0.4	0	2	/	/	/	/	/	/	/	/
FT	0.2	0.4	0	1.5	/	/	/	/	0.8	7.1	0	65.7
G	23.7	29.3	0	98.9	0.8	3.2	0	20.8	2.8	10.5	0	67.6
LT	55.8	33.8	0	100	0.7	2.7	0	15.9	8.6	25	0	100
M	0.2	0.7	0	5.5	0.8	4.9	0	33.7	17.6	35.6	0	100
OB	0.2	0.3	0	1.3	0.4	1.6	0	9.7	0.2	0.5	0	3.3
SM	0.3	1.5	0	9.9	/	/	/	/	0	0.3	0	3.4
UT	10.4	20.2	0	100	96.6	8.3	63.2	100	57.4	44.6	0	100

Note. AH = armed hand; B = blade; FF = front foot; FL = front leg; FT = front thigh; G = guard; LT = lower torso; M = mask; OB = out of bounds; SM = score machine; UT = upper torso.

3.4. Comparisons between Won and Lost Point

3.4.1. Fixation Time

In regard to fixation time (Table 7), on average, fencers during a won point tended to fixate for a longer period of time on the lower torso (mLT = 2159 ms; SD = 3623 ms) and the upper torso (mUT = 2060 ms; SD = 1862 ms) than any other areas. These results were also found in the lost points.

Table 7. Mean, standard deviation, minimum, and maximum average fixation time to particular areas of interest between won and lost point, expressed in ms.

AOI	Point Result							
	Won				Lost			
	Mean	SD	Min	Max	Mean	SD	Min	Max
AH	1347	1570	200	7060	1043	762	268	3622
B	328	133	234	422	559	276	264	985
FF	318	95	217	404	505	201	267	738
FL	236	9	228	246	299	25	281	316
FT	483	344	134	1005	249	77	184	334
G	600	441	167	2145	1204	1975	201	11,094
LT	2159	3623	158	23,451	2661	3309	246	16,132

Table 7. *Cont.*

AOI	Point Result							
	Won				Lost			
	Mean	SD	Min	Max	Mean	SD	Min	Max
M	1191	783	568	3953	1218	1416	211	7649
OB	297	111	200	503	328	306	134	871
SM	244	143	100	387	403	280	167	768
UT	2060	1862	267	11,759	1967	1574	141	7538

Note. AH = armed hand; B = blade; FF = front foot; FL = front leg; FT = front thigh; G = guard; LT = lower torso; M = mask; OB = out of bounds; SM = score machine; UT = upper torso.

3.4.2. Number of Fixations

Focusing on the number of fixations, a contingency table (Table 8) shows a similar distribution between won and lost points, except for lower torso and the armed hand with 1.4 more fixations on it in the lost points. When focusing on the average number of fixations per point (Table 9), we noted that the guard (mG = 1.33; SD = 3.89), the lower torso (mLT = 1.95; SD = 4.69), and the upper torso (mUT = 1.09; SD = 1.54) were the only AOIs that had, on average, one fixation per point, regardless of the result of the point. Lastly, we can note that the armed hand (mAH = 0.91; SD: 2.49) tended to be looked at almost one time, on average, per point in the lost point.

Table 8. Contingency table for number of fixations to particular areas of interest between won and lost points.

AOI	Point Result	
	Won	Loose
AH	83	116
B	2	7
FF	5	6
FL	4	2
FT	6	5
G	183	173
LT	269	194
M	18	34
OB	8	7
SM	4	5
UT	151	154

Note. AH = armed hand; B = blade; FF = front foot; FL = front leg; FT = front thigh; G = guard; LT = lower torso; M = mask; OB = out of bounds; SM = score machine; UT = upper torso.

Table 9. Mean, standard deviation, minimum, and maximum number of fixations to particular areas of interest between won and lost points.

AOI	Point Result							
	Won				Lost			
	Mean	SD	Min	Max	Mean	SD	Min	Max
AH	0.60	2.02	0	12	0.91	2.49	0	16
B	0.01	0.12	0	1	0.06	0.29	0	2

Table 9. *Cont.*

AOI	Point Result							
	Won				Lost			
	Mean	SD	Min	Max	Mean	SD	Min	Max
FF	0.04	0.25	0	2	0.05	0.28	0	2
FL	0.03	0.21	0	2	0.02	0.13	0	1
FT	0.04	0.21	0	1	0.04	0.26	0	2
G	1.33	3.89	0	21	1.35	4.36	0	30
LT	1.95	4.69	0	24	1.52	4.33	0	24
M	0.13	0.40	0	3	0.27	0.61	0	4
OB	0.06	0.29	0	2	0.06	0.29	0	2
SM	0.03	0.21	0	2	0.04	0.23	0	2
UT	1.09	1.54	0	13	1.20	1.60	0	12

Note. AH = armed hand; B = blade; FF = front foot; FL = front leg; FT = front thigh; G = guard; LT = lower torso; M = mask; OB = out of bounds; SM = score machine; UT = upper torso.

3.4.3. Dwell Time (%)

For dwell time (Table 10), whatever the result of the point was, fencers, on average, tended to spend half of the time on the upper torso (mUT = 53.7/53.8; SD = 44.7/45.9). The other half seemed to be devoted to four AOIs: the armed hand (MAH = 11.3/7.4; SD = 25.6/23.2), the guard (mG = 8.5/6.3; SD = 21.1/16), the lower torso (mLT = 14.6/21.6; SD = 30.8/34.2), and the mask (mM = 11.1/9.4; SD = 28.5/28.2).

Table 10. Mean, standard deviation, minimum, and maximum time devoted to particular area of interest between won and lost points, expressed in %.

AOI	Point Result							
	Won				Lost			
	Mean	SD	Min	Max	Mean	SD	Min	Max
AH	11.3	25.6	0	100	7.4	23.2	0	100
B	0.1	0.8	0	6.5	0	0.2	0	2.6
FF	0.1	0.6	0	4.2	0.1	0.4	0	3.5
FL	0	0.1	0	1	0	0.2	0	2
FT	0	0.2	0	1.5	0.9	7.5	0	65.7
G	8.5	21.1	0	98.9	6.3	16	0	75.3
LT	14.6	30.8	0	100	21.6	34.2	0	100
M	11.1	28.5	0	100	9.4	28.2	0	100
OB	0.1	0.7	0	6.3	0.1	0.9	0	9.7
SM	0.2	1.1	0	9.9	0	0.2	0	2.3
UT	53.7	44.7	0	100	53.8	45.9	0	100

Note. AH = armed hand; B = blade; FF = front foot; FL = front leg; FT = front thigh; G = guard; LT = lower torso; M = mask; OB = out of bounds; SM = score machine; UT = upper torso.

4. Discussion

The aim of this exploratory study was to examine the gaze behavior of top-level fencers in ecological settings. More specifically, we wanted to investigate the possibility of various visual search strategies between weapons and between won and lost points. To do so, we examined the visual activity of fencers during a simulated competition. Compared to previous studies [10,11], our results (Tables 4–6) seem to show that the visual activity of foil fencers, in a more ecological situation, meaning without temporal constraints or the requirement to initiate a specific attack or defend against a specific attack, may differ. Also, there may be variations in visual activity among different weapons. Additionally, our study did not identify differences in visual activity between won and lost points in fencing (Tables 7–10). We acknowledge the challenges in drawing conclusions from an exploratory study and the limitations of the results presented here. Nevertheless, the following discussion aims to provide some explanations for the findings in comparison to the existing literature and suggest avenues for future research.

In foil, Witkowski et al. [9] showed that fencers fixated primarily on two AOIs, upper torso and armed hand, and explained this by the fact that attention is directed toward the onset of movement initiation. In our study, foil fencers seemed to fixate for a much longer time (Table 4) and for a higher number of fixations (Table 5) on the upper torso of the opponent in contrast with the other AOIs. Moreover, this AOI was the first and nearly the only one on which a foil fencer fixated during a point, with an observation time of 96.6% (Table 6). This difference in results may be explained by the opponent's handedness, as already demonstrated by Witkowski et al. [10], who showed that experts in front of right-handed opponents fixated primarily on the upper torso with glances on proximal AOIs, like armed hand, guard, or mask. In contrast, in front of a left-handed opponent, experts tended to equally fixate on the upper torso and on the armed hand. In our study, fencers also fought a right-handed opponent; we observed some glances to this aforementioned AOI but with a number of fixations below 0.2 (Table 5) and a dwell time ranging between 0.3% and 0.8% (Table 6). In this study, top-level foil fencers, tier 5 according to McKay et al. [13], anchored their gaze centrally on the upper torso of their opponents and used peripheral vision to react to attacks from the armed hand, like Hausegger et al. [16] showed with martial arts experts or Witkowski et al. [9–11] in fencing. In addition, this AOI can be considered as the gravity center of the foil scoring area [17]. Therefore, fixation on the upper torso enables fencers to monitor the onset of movement initiation (armed hand) and the entire scoring area of the opponent by using foveal and peripheral vision [18].

Secondly, the present study displays some potential differences in visual activity between weapons. Our results indicated that some AOIs were only fixated on one weapon (Tables 4–6), like, for instance, all the AOIs located under the lower torso in epee (i.e., front foot, front leg, front thigh). Furthermore, we noticed that, in contrast with the foil fencer, who mainly fixated on one area, in sabre and epee, an average of three to four areas were fixated per point, with a primary fixation on the upper torso in sabre and then mask, lower torso, and armed hand (Tables 4–6). For epee, we noted a main fixation on the lower torso and then a distribution between guard, upper torso, and armed hand. This difference in visual search strategies can be explained by the inherent rules of practice. In epee, points can be given when a "touche" is executed on a part of the entire body, in contrast with foil, where only the torso can be touched ("touché"), or in sabre, where only the upper part of the body can be touched ("touché") or sliced. Consequently, in addition to monitoring the onset of movement initiation, located on the armed hand, fencers need to track all other areas where they can score. To monitor both of these and reduce the cost associated with saccadic eye movement [19], fencers anchor on a central point (i.e., upper torso in sabre, lower torso in epee) and shift between different cues around this pivot point [17]. This shift between locations can be explained by the importance of those areas for scoring and the need to be processed with the fovea to guarantee the possible movement parameterization and execution of movement to this area (quiet eye; for a review see [20,21]).

Thirdly, the findings appear to indicate that visual activity is not related to a gain or a loss of a point (direct performance indicator) (Tables 7–10). Two hypotheses can be provided to explain these results. The first is that eye tracking enables the collection of data about the activity of foveal vision. It does not characterize the activity of peripheral vision or can only provide an estimate of the peripheral visual field (40° of the visual angle). In addition, during fixation on AOI, two forms of attention may be present. Either the information is processed by foveal vision, in which case, attention and foveal fixation are combined and referred to as overt attention [22], or peripheral vision processes the information, in which case, foveal vision and attentional focus diverge (covert attention) [18]. It is possible that when a fencer looks at a specific area of interest, such as the upper torso, their covert attention may change into overt attention or vice versa, and this could have an impact on the outcome of the point. However, the distinction between these two categories of attention cannot be made based solely on gaze behavior. The second explanation is that the selected participants may not have allowed for the differentiation of visual activity based on performance to be highlighted. Indeed, it is well-established in the scientific literature that, compared to novices, experts exhibit distinct visual activities that underlie different performances [21]. In karate, experts exhibit shorter response time than novices, with fewer fixations of longer duration and fewer locations, compared to novices [23], or in boxing [3], where experts made fewer fixations than novices with fixations mainly directed to the head, whereas novices fixated mainly and longer on the arms and fists, leading to a significant decrement in decision accuracy. However, in this case, with the selected participants (Tier 5 [13]), it is plausible that the difference between points scored and points lost was related to other factors than simple gaze behavior.

Finally, this study, conducted in a simulated competition environment, placed fencers in conditions that aimed to replicate accurately what they may encounter during competitions. Two new areas were identified via the qualitative analysis (AOI) of the visual activity: the score machine and an area referred to as "out of bounds". The score machine was located centrally, next to the piste. The score and, more importantly, the remaining time are displayed in a table. The eye-tracking data analysis revealed that this region was only fixated upon near the end of a match or just before a pause, when there was just a few seconds left. Fixations on this area—which are not the opponent—could be signs of effective time management but can also be risky. Fencers preoccupied with this area stop focusing on their opponent, which can be detrimental. The second emerged area combines a set of points fixated on by the fencer during the bout that do not target a specific area of their opponent. Even though these fixated areas do not carry information, it seems important to include them. It does, in fact, symbolize a particular visual activity—that of directing attention away from an opponent. This category includes fixations on a light box behind the opponent that turns on a specific color after a touch is scored. In this instance, the fixation enables one to confirm whether they have touched or been touched by an opponent. Other fixations in this category are on details that are irrelevant to the game or the opposition. Although these two areas are not directed towards the opponent, they still report a visual activity performed during a competition and potentially reflect what happens in a real match. Therefore, in situ research seems to be more relevant to depicting visual activity as it is performed during a real competition, in contrast to research that depicts visual search strategies in fencing but in controlled situations [9–11] and, therefore, corresponds to a particular situation. It seems important to deliver this information, as it shows that even with experts, top-level athletes, visual activity and, therefore, attention can be disrupted by external elements. These factors should be considered when preparing for major events.

5. Conclusions

Among the limitations of our study that must be taken into account when interpreting these results is the difficulty of generalizing from our extremely small and homogenous sample. We can hypothesize that expanding the sample size for future research would be beneficial. This would allow for a more comprehensive understanding of visual activity in fencing, even revealing weapons-specific differences in visual search strategies. Identifying these differences could lead to the development of distinct training methods and strategies that are tailored to each weapon. Another limitation of this study is related to the use of a remote head-mounted eye tracker device. Although remote eye trackers provide a high degree of ecological validity, they have also been noted to have disadvantages [24,25].

Compared to experimental settings where head-mounted eye trackers and chinrests are used, tracking participants' gaze behavior in less standardized conditions (e.g., when participants' heads are not restrained) may result in a decrease in data quality, particularly in terms of the amount of lost data and precision of the recorded gaze position [26,27]. Moreover, frame-by-frame analysis of this type of data is difficult. First, the margin of error in the accuracy (deviation) provided by this type of eye tracker can make data analysis more difficult. In other words, the actual point of gaze may differ from the estimated point of gaze, which may lead to misunderstanding the fixed AOI. In addition, the distance between the fencer and his or her opponent can modify the size of the AOI, making it larger or smaller depending on the distance. This limitation leads to a requirement to interpret the results with more caution.

Our study focused on the visual activity during a point; however, future research could investigate the impact of decisions made prior to a point on a fencer's visual activity. By examining this relationship, researchers can investigate how a fencer's prior decisions may influence their attentional focus, visual search patterns, and, ultimately, their performance during a match. Additionally, investigating distractor cues and the function of peripheral vision in the management of covert attention will be extremely beneficial. Distractor cues may have a significant effect on an athlete's performance, so it is crucial to understand how to minimize their effects. Furthermore, examining the role of peripheral vision for covert attention could shed light on strategies that can improve an athlete's capacity to process pertinent information while maintaining awareness of their opponents.

These various perspectives could be investigated using mixed-method approaches, such as combining quantitative data obtained from eye-tracking technology with qualitative data gathered through self-confrontation interviews, to investigate these phenomena and attempt to explain the links between attention, decision making, and visual activity [28].

Author Contributions: Conceptualization, P.B., J.F.F., G.M. and M.B.; methodology, P.B., J.F.F. and M.B.; validation, J.F.F., G.M. and M.B.; formal analysis, P.B.; investigation, P.B. and J.F.F.; data curation, P.B.; writing—original draft, P.B.; writing—review and editing, P.B., J.F.F., T.K., C.B., G.K., G.M. and M.B.; visualization, P.B.; supervision, C.B., G.K. and M.B.; project administration, M.B.; funding acquisition, J.F.F., G.M. and M.B. All authors have read and agreed to the published version of the manuscript.

Funding: This study was supported by the French National Research Agency (ANR) with a grant from the Programme d'Investissements d'Avenir [ref. ANR-20-STHP-005].

Institutional Review Board Statement: The study was conducted in accordance with the Declaration of Helsinki, and approved by the Institutional Review Board (or Ethics Committee) of Nantes University (08042021; date: 8 April 2021).

Informed Consent Statement: All participants were informed about the procedures of the study and signed the informed consent.

Data Availability Statement: Data not available.

Acknowledgments: We are extremely grateful to the fencers and coaches on the French national team for their time, effort, and willingness to participate in this study.

Conflicts of Interest: The authors declare no conflict of interest. The funders had no role in the design of the study; in the collection, analyses, or interpretation of data; in the writing of the manuscript; or in the decision to publish the results.

References

1. van Maarseveen, M.J.J.; Savelsbergh, G.J.P.; Oudejans, R.R.D. In Situ Examination of Decision-Making Skills and Gaze Behaviour of Basketball Players. *Hum. Mov. Sci.* **2018**, *57*, 205–216. [CrossRef] [PubMed]
2. de Quel, Ó.M.; Bennett, S.J. Perceptual-Cognitive Expertise in Combat Sports: A Narrative Review and a Model of Perception-Action./Habilidades Perceptivo-Cognitivas En Deportes de Combate: Una Revisión Narrativa y Un Modelo de Percepción-Acción. *RICYDE Rev. Int. Cienc. Deporte* **2019**, *15*, 323–338. [CrossRef]

3. Ripoll, H.; Kerlirzin, Y.; Stein, J.-F.; Reine, B. Analysis of Information Processing, Decision Making, and Visual Strategies in Complex Problem Solving Sport Situations. *Hum. Mov. Sci.* **1995**, *14*, 325–349. [CrossRef]
4. Piras, A.; Pierantozzi, E.; Squatrito, S. Visual Search Strategy in Judo Fighters during the Execution of the First Grip. *Int. J. Sports Sci. Coach.* **2014**, *9*, 185–198. [CrossRef]
5. Roi, G.S.; Bianchedi, D. The Science of Fencing: Implications for Performance and Injury Prevention. *Sports Med.* **2008**, *38*, 465–481. [CrossRef]
6. Silva, A.F.; Afonso, J.; Sampaio, A.; Pimenta, N.; Lima, R.F.; Castro, H.d.O.; Ramirez-Campillo, R.; Teoldo, I.; Sarmento, H.; González Fernández, F.; et al. Differences in Visual Search Behavior between Expert and Novice Team Sports Athletes: A Systematic Review with Meta-Analysis. *Front. Psychol.* **2022**, *13*, 1001066. [CrossRef]
7. Russo, G.; Ottoboni, G. The Perceptual—Cognitive Skills of Combat Sports Athletes: A Systematic Review. *Psychol. Sport Exerc.* **2019**, *44*, 60–78. [CrossRef]
8. Hagemann, N.; Schorer, J.; Cañal-Bruland, R.; Lotz, S.; Strauss, B. Visual Perception in Fencing: Do the Eye Movements of Fencers Represent Their Information Pickup? *Atten. Percept. Psychophys.* **2010**, *72*, 2204–2214. [CrossRef] [PubMed]
9. Witkowski, M.; Tomczak, E.; Bojkowski, Ł.; Borysiuk, Z.; Tomczak, M. Do Expert Fencers Engage the Same Visual Perception Strategies as Beginners? *J. Hum. Kinet.* **2021**, *78*, 49–58. [CrossRef]
10. Witkowski, M.; Tomczak, M.; Bronikowski, M.; Tomczak, E.; Marciniak, M.; Borysiuk, Z. Visual Perception Strategies of Foil Fencers Facing Right- Versus Left-Handed Opponents. *Percept. Mot. Skills* **2018**, *125*, 612–625. [CrossRef]
11. Witkowski, M.; Tomczak, E.; Łuczak, M.; Bronikowski, M.; Tomczak, M. Fighting Left Handers Promotes Different Visual Perceptual Strategies than Right Handers: The Study of Eye Movements of Foil Fencers in Attack and Defence. *BioMed Res. Int.* **2020**, *2020*, 4636271. [CrossRef]
12. Eysenck, M.W.; Derakshan, N.; Santos, R.; Calvo, M.G. Anxiety and Cognitive Performance: Attentional Control Theory. *Emotion* **2007**, *7*, 336–353. [CrossRef]
13. McKay, A.K.A.; Stellingwerff, T.; Smith, E.S.; Martin, D.T.; Mujika, I.; Goosey-Tolfrey, V.L.; Sheppard, J.; Burke, L.M. Defining Training and Performance Caliber: A Participant Classification Framework. *Int. J. Sports Physiol. Perform.* **2022**, *17*, 317–331. [CrossRef]
14. International Fencing Federation. Available online: https://fie.org/athletes (accessed on 20 March 2023).
15. Tonsen, M.; Baumann, C.K.; Dierkes, K. A High-Level Description and Performance Evaluation of Pupil Invisible. *arXiv* **2020**, arXiv:2009.00508. [CrossRef]
16. Hausegger, T.; Vater, C.; Hossner, E.-J. Peripheral Vision in Martial Arts Experts: The Cost-Dependent Anchoring of Gaze. *J. Sport Exerc. Psychol.* **2019**, *41*, 137–145. [CrossRef] [PubMed]
17. Klostermann, A.; Vater, C.; Kredel, R.; Hossner, E.-J. Perception and Action in Sports. On the Functionality of Foveal and Peripheral Vision. *Front. Sports Act. Living* **2020**, *1*, 66. [CrossRef] [PubMed]
18. Williams, A.M.; Ericsson, K.A. Perceptual-Cognitive Expertise in Sport: Some Considerations When Applying the Expert Performance Approach. *Hum. Mov. Sci.* **2005**, *24*, 283–307. [CrossRef]
19. Vater, C.; Williams, A.M.; Hossner, E.-J. What Do We See out of the Corner of Our Eye? The Role of Visual Pivots and Gaze Anchors in Sport. *Int. Rev. Sport Exerc. Psychol.* **2019**, *13*, 81–103. [CrossRef]
20. Lebeau, J.-C.; Liu, S.; Sáenz-Moncaleano, C.; Sanduvete-Chaves, S.; Chacón-Moscoso, S.; Becker, B.J.; Tenenbaum, G. Quiet Eye and Performance in Sport: A Meta-Analysis. *J. Sport Exerc. Psychol.* **2016**, *38*, 441–457. [CrossRef]
21. Mann, D.T.Y.; Williams, A.M.; Ward, P.; Janelle, C.M. Perceptual-Cognitive Expertise in Sport: A Meta-Analysis. *J. Sport Exerc. Psychol.* **2007**, *29*, 457–478. [CrossRef]
22. Piras, A.; Raffi, M.; Perazzolo, M.; Malagoli Lanzoni, I.; Squatrito, S. Microsaccades and Interest Areas during Free-Viewing Sport Task. *J. Sports Sci.* **2019**, *37*, 980–987. [CrossRef] [PubMed]
23. Milazzo, N.; Farrow, D.; Ruffault, A.; Fournier, J.F. Do Karate Fighters Use Situational Probability Information to Improve Decision-Making Performance during on-Mat Tasks? *J. Sports Sci.* **2016**, *34*, 1547–1556. [CrossRef] [PubMed]
24. Kredel, R.; Vater, C.; Klostermann, A.; Hossner, E.-J. Eye-Tracking Technology and the Dynamics of Natural Gaze Behavior in Sports: A Systematic Review of 40 Years of Research. *Front. Psychol.* **2017**, *8*, 1845. [CrossRef] [PubMed]
25. Kredel, R.; Hernandez, J.; Hossner, E.-J.; Zahno, S. Eye-Tracking Technology and the Dynamics of Natural Gaze Behavior in Sports: An Update 2016–2022. *Front. Psychol.* **2023**, *14*, 1130051. [CrossRef] [PubMed]
26. Hooge, I.T.C.; Niehorster, D.C.; Hessels, R.S.; Benjamins, J.S.; Nyström, M. How Robust Are Wearable Eye Trackers to Slow and Fast Head and Body Movements? *Behav. Res.* **2022**, 1–15. [CrossRef] [PubMed]
27. Niehorster, D.C.; Santini, T.; Hessels, R.S.; Hooge, I.T.C.; Kasneci, E.; Nyström, M. The Impact of Slippage on the Data Quality of Head-Worn Eye Trackers. *Behav. Res.* **2020**, *52*, 1140–1160. [CrossRef]
28. Bossard, C.; Kérivel, T.; Dugény, S.; Bagot, P.; Fontaine, T.; Kermarrec, G. Naturalistic Decision-Making in Sport: How Current Advances Into Recognition Primed Decision Model Offer Insights for Future Research in Sport Settings? *Front. Psychol.* **2022**, *13*, 936140. [CrossRef]

Journal of
Functional Morphology and Kinesiology

MDPI

Article

Accumulated Workload Differences in Collegiate Women's Soccer: Starters versus Substitutes

Maxine Furtado Mesa, Jeffrey R. Stout, Michael J. Redd and David H. Fukuda *

School of Kinesiology and Rehabilitation Sciences, University of Central Florida, Orlando, FL 32816, USA; maxine.furtado@ucf.edu (M.F.M.); jeffrey.stout@ucf.edu (J.R.S.); redd@ucf.edu (M.J.R.)
* Correspondence: david.fukuda@ucf.edu

Abstract: The purpose of this study was to estimate the workloads accumulated by collegiate female soccer players during a competitive season and to compare the workloads of starters and substitutes. Data from 19 college soccer players (height: 1.58 ± 0.06 m; body mass: 61.57 ± 6.88 kg) were extracted from global positioning system (GPS)/heart rate (HR) monitoring sensors to quantify workload throughout the 2019 competitive season. Total distance, distance covered in four speed zones, accelerations, and time spent in five HR zones were examined as accumulated values for training sessions, matches, and the entire season. Repeated-measures ANOVA and Student's t tests were used to determine the level of differences between starter and substitute workloads. Seasonal accumulated total distance ($p < 0.001$), sprints (≥19.00 km/h; $p < 0.001$), and high-speed distance (≥15.00 km/h; $p = 0.005$) were significantly greater for starters than substitutes. Accumulated training load ($p = 0.08$) and training load per minute played in matches ($p = 0.08$) did not differ between starters and substitutes. Substitutes had similar accumulated workload profiles during training sessions but differed in matches from starters. Coaches and practitioners should pursue strategies to monitor the differences in workload between starters and substitutes.

Keywords: technology; athlete monitoring; player tracking; football; non-starters

Citation: Furtado Mesa, M.; Stout, J.R.; Redd, M.J.; Fukuda, D.H. Accumulated Workload Differences in Collegiate Women's Soccer: Starters versus Substitutes. *J. Funct. Morphol. Kinesiol.* **2023**, *8*, 78. https://doi.org/10.3390/jfmk8020078

Academic Editors: Diego Minciacchi and Helmi Chaabene

Received: 26 April 2023
Revised: 1 June 2023
Accepted: 9 June 2023
Published: 12 June 2023

1. Introduction

Athlete monitoring examines the physiological stress placed on the body due to physical activity, also known as training load [1]. Training load can be measured internally, reflected as a psychophysiological response to physical activity, or externally, reflected as the physical work performed by the body [2]. Tracking training load variables through the use of global positioning systems (GPSs) has become a common practice in collegiate soccer [3–5], with research being increasingly conducted on female collegiate soccer athletes in recent years [4,6–9]. Monitoring training load could be helpful in detecting changes in fatigue levels during competitive periods when extensive physical performance evaluations are not practical [10]. In addition to detecting variations in fatigue, monitoring training load may help maximize the physical potential of each athlete through individualized approaches to training and recovery.

Substitutions have an important influence on the tactical considerations of a coach's game plan. In collegiate soccer, an unlimited number of substitutions can be made in the second half of a match, whereas in the first half, an athlete who is substituted out must wait for the second half to be substituted back in [11]. Previous research on elite male soccer players demonstrated that substitutes showed higher work rates than the players they substituted in a competitive match [12,13]. Although no differences in successful pass percentages were observed between elite male soccer substitutes and starters, substituted players covered more distance at a high intensity (≥19.8 km/h; 12.4 ± 5.3 m/min) than players who participated in the entire match (9.8 ± 3.2 m/min) or were substituted (11.3 ± 3.2 m/min) due to the replacement of fatigued players or the need for tactical disruptions [14–16]. Similarly, decreases in exercise intensity were reduced with the incorporation of substitutes during matches [17].

The number of minutes played by each athlete during soccer matches also affects the physiological preparation in both sexes [6,12]. Although there are data available on the demands of elite female soccer [6,8,9,18–22], there is a larger representation of research on the demands of elite and collegiate male soccer players [11,12,16,17,23–26]. Additionally, research in the elite female soccer population may not be applicable to the collegiate female soccer population because there are differences in the demands of matches between collegiate and elite female soccer [20,21,27,28]. For example, elite female soccer players cover an average of 10 km per match [18,29,30], while female collegiate soccer players cover less than 10 km and at lower speed thresholds (<15.6 km/h) [31]. When accounting for variations in training strain and training monotony, no significant differences were found between elite female soccer starters and substitutes [22]. Furthermore, elite female starters produced higher maximal running velocities and aerobic capacities than their non-starting or substitute teammates [6]. Conversely, no significant differences were observed in sprint time or submaximal exercise tests between starting and substitute collegiate athletes. However, worthwhile differences were observed when the starters achieved faster 30 m sprint times than substitutes [32]. Furthermore, substitute collegiate soccer players engaged in greater distances of moderate-intensity running (12.1–15.5 km/h) in matches than starting collegiate soccer players [31]. Collegiate female soccer players also experience higher training loads and a decrease in power output during the season [33]. Based on the existing literature, there are conflicting results between starting and substitute players regarding in-game performance indicators, thus emphasizing the importance of identifying the underlying factors behind monitoring workloads.

Substitute players sometimes display better technical qualities than the players on the field for the full game or the players who were replaced [34]. It is important that substitute players maintain fitness and skill throughout the season to match the high loads per minute they experience when entering the game at a later point. Due to a short and congested college season, teams must maximize roster availability by maintaining the fitness of all players. Tracking workload can allow coaches to prescribe 'top-up' conditioning sessions for players who do not receive enough training stimuli during the week [25]. Therefore, it is important to be aware of the workloads imposed on starters compared to substitutes to monitor and adapt training sessions. This study aimed to estimate the workloads accumulated by collegiate female soccer players during a competitive season and compare the workloads of the starting players with those of the substitutes on a collegiate female soccer team. The authors hypothesize that the workloads accumulated during one competitive season will be higher in starters than substitutes.

2. Materials and Methods

This study was a retrospective observational study conducted on one National Collegiate Athletics Association (NCAA, Indianapolis, IN, USA) Division I women's soccer team over the course of one competitive season (August 2019–November 2019). Workload data were only provided for on-field training and competitive matches. A total of 1245 match and training player sessions were used to generate the seasonal accumulated data for the 19 collegiate women's soccer players evaluated for this study. Match data included 311 data files, while training data included 934 data files for starters and substitutes. For this study, seasonal accumulated data were defined as the summation of all on-field training sessions and matches. Thirteen regulation time and five overtime matches were included, with 139 data files from starters and 172 data files from substitutes. Starters averaged 78 ± 13.66 min per game, while substitutes averaged 36 ± 13.92 min per game. Complete data were available for 54 training sessions, averaging 70 ± 25.36 min per training session, with 381 data files belonging to starters and 553 data files for substitutes.

2.1. Subjects

Nineteen NCAA Division I college women's soccer players (age: 20 ± 1.61 years; height: 1.58 ± 0.06 m; body mass: 61.57 ± 6.88 kg) were included in the analysis. To be included in analysis, athletes had to have been medically cleared to participate in the women's soccer team's training sessions and competitive matches and remained healthy with no injuries throughout the season analyzed. Athletes also had to comply with wearing the GPS/heart rate monitor, which included wearing the Polar Team Pro electrode strap on the xiphoid process for the entire duration of the activity. To determine the status of the starters (n = 8) versus the substitutes (n = 11), a limit of >50% of the total match time for the entirety of the season was used based on previous research [8,35]. Status of starters versus substitutes was held constant for match and training data analysis. The study was conducted in accordance with the Declaration of Helsinki and approved by the Institutional Review Board at the University of Central Florida (IRB #2763) on the 22nd of February in 2021. The current study was a retrospective observational study approved by the team's coaching staff. Informed consent was obtained from all subjects involved in the study.

2.2. Procedures

Athletes were assigned individual GPS/heart rate monitors (Polar Team Pro, Polar Electro, Oy, Finland) and chest straps prior to the start of the season as part of team monitoring during all training sessions and matches. The Polar Team Pro sensors have previously been deemed valid and reliable for total distance, low-speed running (0–13.99 km/h), high-speed running (14–19.99 km/h), and very high speed running (>20 km/h) in the outdoor setting [36]. To prevent inter-unit error, athletes wore the same sensor for all training sessions and matches [36,37]. Athletes were given their respective sensors to clip onto a chest strap with electrodes attached once stepping onto the training or match pitch. Sensor placement on the chest strap was located on the xiphoid process and athletes were instructed to ensure the electrodes on the inside of the chest strap made full contact with skin. Sensors would turn on as soon as contact between skin and electrodes on the chest strap was made. Data collection began as soon as field activity was started, including warm-up, and was concluded as soon as the sport coach stopped a training session or the final whistle was blown by the match official. All sessions were recorded live on an iPad (Apple, Cupertino, CA, USA) with the Polar Team Pro app(version 2.0.4). After the activity stopped, the sensors were collected and placed on the dock to be imported into the Polar Team Pro online database. Data were exported from the dashboard to Microsoft Excel spreadsheets (Excel 2019, Microsoft Corporation, Redmond, WA, USA) for analysis.

Specific heart rate zones were used to quantify intensity [8,35] and defined by the default of Polar Team Pro as zone 1 = 50–60%, zone 2 = 60–70%, zone 3 = 70–80%, zone 4 = 80–90%, and zone 5 = 90–100%. Heart rate zones were calculated using the maximal heart rate obtained from a Yo-Yo Intermittent Recovery Test Level 1 (YYIRT) completed at the beginning of the pre-season. Training load, taken directly from Polar Team Pro default, was defined in arbitrary units as the amount of effort put into a session based on intensity and duration. The intensity of a session was determined by proprietary algorithms developed by Polar, including training history, weight, VO2max, sex, age, and heart rate. A count of the frequency of accelerations was quantified into three previously established thresholds [8,35], as follows: low = 0.5–1.99 m/s^2, moderate = 2.00–2.99 m/s^2, and high = 3.00–5.00 m/s^2. Speed zones were separated into four previously established groups with the following thresholds: walk/stand $\leq$ 6.99 km/h, jog = 7.0–14.99 km/h, run 15.0–18.99 km/h, and sprint $\geq$ 19.00 km/h [8]. The run and sprint speed zones ($\geq$15.00 km/h) were combined to define the high-speed distance (HSD), as specified in Jagim et al. [8].

2.3. Statistical Analysis

Normal distribution was established through visual inspection of normal Q-Q plots. Student's t tests were used to examine distance metrics, sprints, and training load differences between starters and substitutes. A two-way repeated-measures analysis (zone × group) repeated-measures analysis of variance was used to examine the movement characteristics between starters and substitutes. Bonferroni post hoc analysis was used to determine where there were differences when significant main effects were identified. To calculate the level of differences in workload, effect sizes were calculated. Effect sizes were interpreted as follows: 0.2 (trivial), 0.2–0.6 (small), 0.7–1.2 (moderate), 1.3–2.0 (large), >2.0 (very large) [38]. Data are presented as mean ± standard deviation. Statistical software (JASP, V.16, Amsterdam, The Netherlands) was used for all analyses. Statistical significance was set a priori at $p < 0.05$.

3. Results

Seasonal accumulated total distance was significantly higher for starters (337.76 ± 26.28 km) compared to substitutes (246.37 ± 39.01 km; t[17] = 5.72, $p < 0.001$) and matches (starters: 201.58 ± 19.82 km vs. substitutes: 107.09 ± 40.65 km; t[17] = 6.69, $p < 0.001$), as shown in Figure 1. Training load during matches was significantly greater for starters (4586.00 ± 1488.32 a.u.) compared to substitutes (2501.73 ± 1150.54 a.u.; t[17] = 3.45, $p = 0.003$), as shown in Figure 2. There was no significant difference in average training load per minute played in matches between starters (3.15 ± 1.12 a.u./minute) compared to substitutes (4.87 ± 2.42 a.u./minute; t[17] = −1.86, $p = 0.08$).

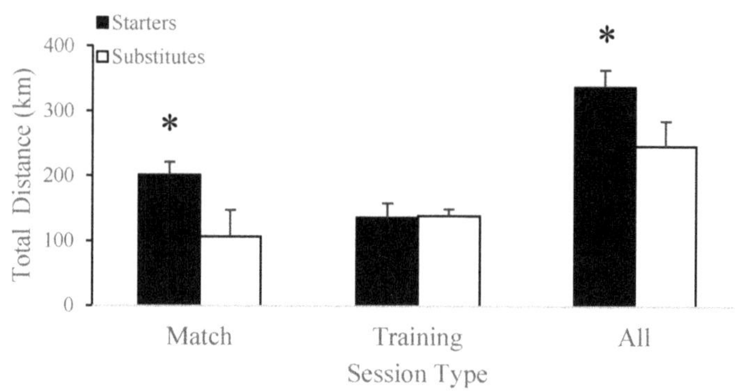

Figure 1. Differences in total distance covered between starters and substitutes in matches, training sessions, and the accumulated competitive season. * $p < 0.05$.

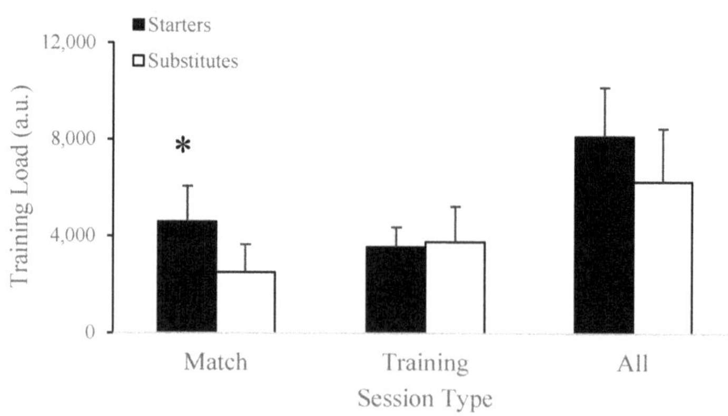

Figure 2. Differences in training load between starters and substitutes in matches, training sessions, and the accumulated competitive season. * $p < 0.05$.

Seasonal accumulated sprints covered were significantly greater for starters (8169.63 ± 440.85 sprints) compared to substitutes (5771.55 ± 906.55 sprints; t[17] = 6.88, $p < 0.001$) and matches (starters: 4879.88 ± 485.43 sprints vs. substitutes: 2363.27 ± 1040.96 sprints; t[17] = 7.07, $p < 0.001$), as shown in Figure 3. Seasonal accumulated high-speed distance was significantly higher for starters (36.87 ± 6.57 km) compared to substitutes (25.55 ± 8.10 km; t[17] = 3.23, $p = 0.005$) and matches (starters: 24.70 ± 5.12 km vs. substitutes: 12.83 ± 6.92 km; t[17] = 4.09, $p < 0.001$), as shown in Figure 4. Seasonal accumulated training load did not differ between starters (8130.88 ± 2026.60 a.u.) and substitutes (6261.00 ± 2201.22 a.u.; t[17] = 1.89, $p = 0.08$).

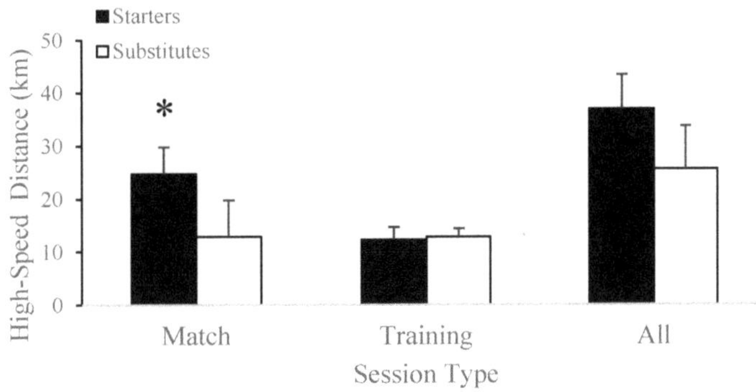

Figure 3. Differences in high-speed distance covered between starters and substitutes in matches, training sessions, and the accumulated competitive season. * $p < 0.05$.

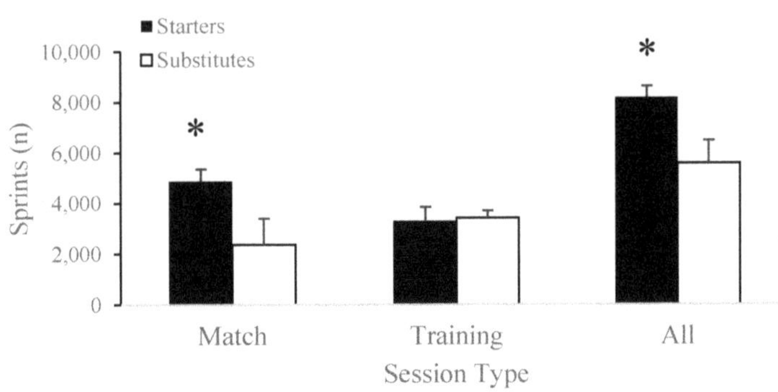

Figure 4. Differences in total count of sprints covered between starters and substitutes in matches, training sessions, and the accumulated competitive season. * $p < 0.05$.

Total distance (starters: 136.17 ± 22.08 km vs. substitutes: 139.28 ± 9.79 km; t[17] = −0.42, $p = 0.68$), training load (starters: 3544.88 ± 812.05 a.u. vs. substitutes: 3759.27 ± 1480.45 a.u.; t[16.06] = 0.40, $p = 0.69$), sprints (starters: 3289.75 ± 551.14 sprints vs. substitutes: 3408.27 ± 292.41 sprints; t[17] = −0.61, $p = 0.55$), and high-speed distance (starters: 12.17 ± 2.47 km vs. substitutes: 12.72 ± 1.68 km; t[11.58] = −0.55, $p = 0.60$) did not differ between starters and substitutes during training sessions throughout the competitive season.

Tables 1–3 show the accumulated workloads by zone for matches only, training sessions only, and all sessions, respectively. Speed zones, heart rate zones, and acceleration zone metrics are displayed in accumulated kilometers, minutes, and counts, respectively. A repeated-measures analysis of variance (ANOVA) on speed zones resulted in a significant speed zone × player status interaction in mean differences between groups,

$F(1.584, 26.924) = 6.203$, $p = 0.01$. A post hoc Bonferroni test showed significantly different seasonal accumulated distances between starters and substitutes in speed zone 1 ($p = 0.002$, d = 1.91) and speed zone 2 ($p < 0.001$, d = 1.33). A repeated-measures ANOVA on speed zones during matches resulted in a significant speed zone × player status interaction in mean differences between groups, $F(1.282, 21.789) = 7.498$, $p = 0.008$. A post hoc Bonferroni test showed significantly different match distance covered between starters and substitutes in speed zone 1 ($p = 0.001$, d = 2.53) and speed zone 2 ($p < 0.001$, d = 1.30). A repeated-measures ANOVA on seasonal accumulated and match acceleration counts in different acceleration zones resulted in a significant speed zone × player status interaction in mean differences between groups, $F(1.006, 17.106) = 14.064$, $p = 0.002$ and $F(1.009, 17.147) = 40.243$, $p < 0.001$, respectively. A post hoc Bonferroni test showed significantly different seasonal accumulated counts of high-zone accelerations between starters and substitutes ($p < 0.001$, d = 1.85). A post hoc Bonferroni test showed significantly different counts of high-zone accelerations during matches between starters and substitutes ($p < 0.001$, d = 3.19).

Table 1. Seasonal accumulated workloads for matches and training sessions combined in collegiate Division I women's soccer players by starting status (mean ± SD).

Variable	Starters	Subs	*p*	Cohen's d
SP1 (km)	128.20 ± 21.05	94.19 ± 13.94	0.002	1.91
SP2 (km)	149.95 ± 38.52	111.29 ± 13.94	<0.001	1.33
SP3 (km)	22.25 ± 3.28	15.74 ± 4.85	1.000	1.57
SP4 (km)	13.62 ± 4.02	9.81 ± 3.98	1.000	0.95
HR1 (min)	1314.71 ± 451.11	1374.68 ± 401.45	1.000	−0.14
HR2 (min)	1439.75 ± 308.32	1385.06 ± 252.58	1.000	0.19
HR3 (min)	1183.69 ± 204.19	959.23 ± 289.54	1.000	0.90
HR4 (min)	1007.28 ± 420.63	724.15 ± 407.97	1.000	0.68
HR5 (min)	438.42 ± 508.66	262.29 ± 287.59	1.000	0.43
AZ1 (n)	39,203.13 ± 3982.65	31,914.00 ± 3881.33	<0.001	1.85
AZ2 (n)	3480.75 ± 308.05	2530.73 ± 401.91	1.000	2.65
AZ3 (n)	922.25 ± 200.45	654.00 ± 182.24	1.000	1.40

Speed zones: SP1 (walk/stand) ≤ 6.99 km/h; SP2 (jog) = 7.00–14.99 km/h; SP3 (run) = 15.00–18.99 km/h; SP4 (sprint) ≥ 19.00 km/h. Heart rate zones: HR1 = 50–60%; HR2 = 60–70%; HR3 = 70–80%, HR4 = 80–90%; HR5 = 90–100%. Acceleration zones: AZ1 (low) = 0.5–1.99 m/s^2; AZ2 (moderate) = 2.00–2.99 m/s^2; AZ3 (high) = 3.00–5.00 m/s^2.

Table 2. Total accumulated match workloads in collegiate DI women's soccer players by starting status (mean ± SD).

Variable	Starters	Subs	*p*	Cohen's d
SP1 (km)	75.39 ± 14.05	40.88 ± 13.20	0.001	2.53
SP2 (km)	101.02 ± 38.72	59.33 ± 23.51	<0.001	1.30
SP3 (km)	16.01 ± 2.63	8.20 ± 4.44	1.000	2.14
SP4 (km)	8.69 ± 2.98	4.63 ± 2.98	1.000	1.36
HR1 (min)	488.28 ± 273.00	581.35 ± 290.02	1.000	−0.33
HR2 (min)	586.96 ± 146.45	514.66 ± 152.88	1.000	0.48
HR3 (min)	584.78 ± 146.26	354.72 ± 112.46	1.000	1.76
HR4 (min)	637.05 ± 251.46	317.62 ± 256.91	0.187	1.26
HR5 (min)	348.33 ± 421.30	112.59 ± 161.85	1.000	0.74
AZ1 (n)	20,021.63 ± 1367.71	12,337.73 ± 3124.69	<0.001	3.19
AZ2 (n)	1901.38 ± 199.73	957.91 ± 400.10	1.000	2.98
AZ3 (n)	456.88 ± 93.15	246.55 ± 116.15	1.000	2.00

Speed zones: SP1 (walk/stand) ≤ 6.99 km/h; SP2 (Jog) = 7.00–14.99 km/h; SP3 (run) = 15.00–18.99 km/h; SP4 (sprint) ≥ 19.00 km/h. Heart rate zones: HR1 = 50–60%; HR2 = 60–70%; HR3 = 70–80%, HR4 = 80–90%; HR5 = 90–100%. Acceleration zones: AZ1 (low) = 0.5–1.99 m/s^2; AZ2 (moderate) = 2.00–2.99 m/s^2; AZ3 (high) = 3.00–5.00 m/s^2.

Table 3. Total accumulated training workloads in collegiate DI women's soccer players by starting status (mean ± SD).

Variable	Starters	Subs	*p*	Cohen's d
SP1 (km)	53.51 ± 10.34	53.31 ± 5.88	1.000	0.02
SP2 (km)	48.93 ± 8.20	51.96 ± 5.13	1.000	−0.44
SP3 (km)	7.24 ± 1.39	7.54 ± 0.71	1.000	−0.27
SP4 (km)	4.93 ± 1.31	5.18 ± 1.13	1.000	−0.20
HR1 (min)	826.43 ± 205.33	767.46 ± 215.67	1.000	0.28
HR2 (min)	852.80 ± 188.94	841.87 ± 185.58	1.000	0.06
HR3 (min)	598.91 ± 160.62	599.46 ± 229.99	1.000	−0.002
HR4 (min)	370.23 ± 192.78	399.89 ± 251.78	1.000	−0.13
HR5 (min)	90.09 ± 88.55	138.61 ± 155.54	1.000	−0.38
AZ1 (n)	19,181.50 ± 3746.08	19,576.27 ± 2294.86	1.000	−0.13
AZ2 (n)	1579.38 ± 307.42	1572.82 ± 207.11	1.000	0.003
AZ3 (n)	465.38 ± 118.70	407.46 ± 80.44	1.000	0.57

Speed zones: SP1 (walk/stand) ≤ 6.99 km/h; SP2 (Jog) = 7.00–14.99 km/h; SP3 (run) = 15.00–18.99 km/h; SP4 (sprint) ≥ 19.00 km/h. Heart rate zones: HR1 = 50–60%; HR2 = 60–70%; HR3 = 70–80%, HR4 = 80–90%; HR5 = 90–100%. Acceleration zones: AZ1 (low) = 0.5–1.99 m/s^2; AZ2 (moderate) = 2.00–2.99 m/s^2; AZ3 (high) = 3.00–5.00 m/s^2.

4. Discussion

The objective of this study was to estimate the workloads accrued by collegiate women's soccer players over the course of the competitive season, which included all games and practices. The workloads of the starters and substitutes for the same squad were compared as a secondary goal. A key finding of the study was the discrepancy between starters and substitutes in overall match workloads. Discrepancies between starters and substitutes were expected based on previous studies [8,23].

Comparing work rates and absolute values of starting and substitute soccer players becomes complex, as the definitions and time limits considered to define non-starting and substitution players vary. Varying time limits for playing status separation complicates the comparison of results between studies. To separate playing status for analysis, a few studies set a minimum threshold for minutes played. For example, Carling et al. [12] characterized those who played a minimum of 10 min per game. Similarly, Gai et al. [26] and Hills et al. [25] specified a minimum playing time of five minutes for inclusion in analysis. Gimenez et al. [24] considered substitutes as players who played less than 65 min per match during the regular season. Other studies separated playing status by percentages. For example, Curtis et al. [23] included players as starters if they started in more than 60% of the total matches in the season. The methodology of this current study was based on that of Jagim et al., where substitutes were considered to be those that played less than 50% of the total match time [8]. Lorenzo-Martinez et al. [34] did not specify playing time, but excluded substitutions made in the first half and during stoppage time.

The substitutes in the current study covered significantly lower total distances (31% average difference), high-speed distances (63% average difference) and numbers of sprints (34% average difference) than starters. Percentages of average difference for significant values were obtained through group averages for starters and substitutes. Similar results were observed in collegiate female soccer players playing in the third division, where starters had significantly greater values of total distance, high-speed distance, training load, and number of sprints during matches and for seasonal accumulated values than substitutes; however, no differences were noted in training sessions [8]. In contrast, elite substitute soccer players likely covered more absolute high-intensity running distances at >4.2 ≤ 5 m/s and >5 ≤ 6.9 m/s (30.5% average difference), and had higher player loads (13.9% average difference), which is calculated differently than the currently examined training load, compared to starting players [24]. The contrast in results may be due to the use of two friendly matches in the previous study versus an entire competitive season in the current study. When considering work rate relative to minutes played in professional male soccer during friendly matches, substitute players covered higher total distances

(4.6% average difference compared to starting players) [39]. Previous research reports that playing time can potentially influence running performance indices, such as differences in cadence in the game or pacing strategies during the time spent on the field [40].

The current study did not demonstrate significant differences in external load markers such as total and high-speed distance covered, number of sprints, or training load during training sessions between substitutes and starters (Table 3, Figure 1A,B and Figure 2A,B). It is important to note that despite the discrepancy between the results of the current study and those of the existing literature, significant variation exists between substitution rules in professional and American collegiate soccer. Although professional soccer coaches are only allowed a total of five substitutes during the whole game, collegiate soccer coaches are allowed unlimited substitutions in the second half. Because 60% of collegiate men's soccer substitutions count as re-substitutions, the workload of substitutes is lower than that of starters during matches because of their limited participation [23]. Supporting this, Vescovi and Favero [31] demonstrated that collegiate female soccer substitutes covered shorter distances at moderate (15% vs. 19%) and high intensity (6% vs. 16%) than the starting players in competitive matches. Comparable results were also found in male soccer substitutes, who showed significantly less heart-rate-weighted training impulse, total distance, and acceleration counts than starters during a competitive season [23]. Furthermore, imposing high loads on low-minute players may put those athletes at a higher risk of injury than players with higher minutes [41]. Similar to the increased risk of injury during weeks of highly loaded preseason training sessions [42], consistently exposing substitute players to higher loads during training sessions while they continue to experience lower loads during matches may pose higher risks of injury. Therefore, it may be of interest for teams to track training loads separately for starters and substitutes throughout a season to monitor for discrepancies.

Accelerations account for 7–10% of the total training load during competitive matches [43], and an increase in weekly accelerations can increase fatigue throughout a competitive season [19]. Acceleration counts provide a more comprehensive understanding of the amount of energy expended during a match [44], allowing a more detailed approach to the physical workload experienced by players during activity. However, significant differences were only observed for matches and seasonal accumulated low-zone accelerations, where starters performed more match and seasonal accumulated accelerations (Table 1). To prevent large spikes in workload for substitutes when trying to fill any missed load during matches, coaches may be able to recreate similar acceleration patterns during small-sided games during training [45]. Sport coaches may opt for varied small-sided game dimensions to elicit the preferred adaptations depending on the training day and its proximity to a match day.

There are a few limitations of the current study that should be noted. The current study only examined physical workloads obtained through a GPS/heart rate sensor. Physical characteristics and workloads do not address the complete picture throughout a competitive season, as sport coaches consider tactical and technical skills when deciding the starting roster and substitutions. Future research should include tactical variables such as pass completion and ball possession to further determine if any differences exist between starters and substitutes. Furthermore, the integrity of the dataset was maintained through a smaller sample size due to the exclusion of non-compliant sensor-wearing athletes by thorough data analysis. Of the 30 members on the team, 36% of the data were unusable due to injury, no GPS/heart rate sensor assignment, or non-compliance. One-third of the unusable data was due to non-compliance of sensor wear. Additionally, the current study did not normalize all data for playing time as we looked at values in an accumulated manner. Although previous studies have compared starters' and substitutes' values in an absolute manner [8,23,26,46–50], future research may look to replicate this study with external load variables relative to playing time [16,39,51–54].

Quantifying workloads allows you to see the physical stress that starters and substitutes face throughout a season. The results of this study indicate the differences in workload between starting and substitute soccer players with varying minutes of activity.

Coaches and practitioners should strive to implement strategies to monitor the differences in physical workload between starters and substitutes.

5. Conclusions

The results of this study show differences in the seasonal accumulated and match workloads between starters and substitutes on a college women's soccer team. Starters showed significantly higher accumulated total distance, sprints, and high-speed distance throughout a competitive season and significantly higher absolute total distance, sprints, high-speed distance, and training load during competitive matches. Despite workload differences in matches, no significant differences were observed between starters and substitutes during training sessions.

Author Contributions: Conceptualization, M.F.M., J.R.S., D.H.F. and M.J.R.; methodology, M.F.M., J.R.S. and D.H.F.; formal analysis, M.F.M., J.R.S. and D.H.F.; investigation, M.F.M.; resources, M.F.M., J.R.S., D.H.F. and M.J.R.; data curation, M.F.M. and J.R.S.; writing—original draft preparation, M.F.M., J.R.S. and D.H.F.; writing—review and editing, M.F.M., J.R.S., D.H.F. and M.J.R.; supervision, J.R.S. All authors have read and agreed to the published version of the manuscript.

Funding: This research received no external funding.

Institutional Review Board Statement: The study was conducted in accordance with the Declaration of Helsinki, and approved by the Institutional Review Board at the University of Central Florida (IRB #2763) approved 22 February 2021.

Informed Consent Statement: Informed consent was obtained from all subjects involved in the study.

Data Availability Statement: The data presented in this study are available on request from the corresponding author.

Conflicts of Interest: The authors declare no conflict of interest.

References

1. Impellizzeri, F.M.; Rampinini, E.; Marcora, S.M. Physiological Assessment of Aerobic Training in Soccer. *J. Sports Sci.* **2005**, *23*, 583–592. [CrossRef]
2. Impellizzeri, F.M.; Marcora, S.M.; Coutts, A.J. Internal and External Training Load: 15 Years on. *Int. J. Sports Physiol. Perform* **2019**, *14*, 270–273. [CrossRef] [PubMed]
3. Huggins, R.A.; Fortunati, A.R.; Curtis, R.M.; Looney, D.P.; West, C.A.; Lee, E.C.; Fragala, M.S.; Hall, M.L.; Casa, D.J. Monitoring Blood Biomarkers and Training Load throughout a Collegiate Soccer Season. *J. Strength Cond. Res.* **2019**, *33*, 3065–3077. [CrossRef]
4. McFadden, B.A.; Walker, A.J.; Bozzini, B.N.; Sanders, D.J.; Arent, S.M. Comparison of Internal and External Training Loads in Male and Female Collegiate Soccer Players during Practices vs. Games. *J. Strength Cond. Res.* **2020**, *34*, 969–974. [CrossRef]
5. Slater, L.V.; Baker, R.; Weltman, A.L.; Hertel, J.; Saliba, S.A.; Hart, J.M. Activity Monitoring in Men's College Soccer: A Single Season Longitudinal Study. *Res. Sports Med.* **2018**, *26*, 178–190. [CrossRef] [PubMed]
6. Manson, S.A.; Brughelli, M.; Harris, N.K. Physiological Characteristics of International Female Soccer Players. *J. Strength Cond. Res.* **2014**, *28*, 308–318. [CrossRef] [PubMed]
7. Kupperman, N.; DeJong, A.F.; Alston, P.; Hertel, J.; Saliba, S.A. Athlete Workloads During Collegiate Women's Soccer Practice: Implications for Return to Play. *J. Athl. Train.* **2020**, *56*, 321–330. [CrossRef] [PubMed]
8. Jagim, A.R.; Askow, A.T.; Carvalho, V.; Murphy, J.; Luedke, J.A.; Erickson, J.L. Seasonal Accumulated Workloads in Collegiate Women's Soccer: A Comparison of Starters and Reserves. *J. Funct. Morphol. Kinesiol.* **2022**, *7*, 11. [CrossRef] [PubMed]
9. Bozzini, B.N.; McFadden, B.A.; Walker, A.J.; Arent, S.M. Varying Demands and Quality of Play between In-Conference and Out-of-Conference Games in Division I Collegiate Women's Soccer. *J. Strength Cond. Res.* **2020**, *34*, 3364–3368. [CrossRef] [PubMed]
10. Thorpe, R.T.; Atkinson, G.; Drust, B.; Gregson, W. Monitoring Fatigue Status in Elite Team Sport Athletes: Implications for Practice. *Int. J. Sports Physiol. Perform.* **2017**, *12*, S2–S27. [CrossRef] [PubMed]
11. Conte, D.; Niederhausen, M.; LaPlante, M.; Tessitore, A. Substitution Patterns and Analysis in Men's Division I College Soccer. In *International Research in Science and Soccer II*; Routledge: Oxforshire, UK, 2015; pp. 44–52. ISBN 1-315-68674-0.
12. Carling, C.; Espié, V.; Le Gall, F.; Bloomfield, J.; Jullien, H. Work-Rate of Substitutes in Elite Soccer: A Preliminary Study. *J. Sci. Med. Sport* **2010**, *13*, 253–255. [CrossRef]
13. Mohr, M.; Krustrup, P.; Bangsbo, J. Match Performance of High-Standard Soccer Players with Special Reference to Development of Fatigue. *J. Sports Sci.* **2003**, *21*, 519–528. [CrossRef]
14. Reilly, T.; Drust, B.; Clarke, N. Muscle Fatigue during Football Match-Play. *Sports Med.* **2008**, *38*, 357–367. [CrossRef] [PubMed]

15. Siegle, M.; Lames, M. Game Interruptions in Elite Soccer. *J. Sports Sci.* **2012**, *30*, 619–624. [CrossRef]
16. Bradley, P.S.; Lago-Peñas, C.; Rey, E. Evaluation of the Match Performances of Substitution Players in Elite Soccer. *Int. J. Sports Physiol. Perform* **2014**, *9*, 415–424. [CrossRef]
17. Coelho, D.B.; Coelho, L.G.M.; Morandi, R.F.; Ferreira Junior, J.B.; Marins, J.C.B.; Prado, L.S.; Soares, D.D.; Silami-Garcia, E. Effect of Player Substitutions on the Intensity of Second-Half Soccer Match Play. *Rev. Bras. Cineantropometria Desempenho Hum.* **2012**, *14*, 183–191.
18. Krustrup, P.; Mohr, M.; Ellingsgaard, H.; Bangsbo, J. Physical Demands during an Elite Female Soccer Game: Importance of Training Status. *Med. Sci. Sport. Exerc.* **2005**, *37*, 1242–1248. [CrossRef] [PubMed]
19. Douchet, T.; Humbertclaude, A.; Cometti, C.; Paizis, C.; Babault, N. Quantifying Accelerations and Decelerations in Elite Women Soccer Players during Regular In-Season Training as an Index of Training Load. *Sports* **2021**, *9*, 109. [CrossRef] [PubMed]
20. Datson, N.; Drust, B.; Weston, M.; Jarman, I.H.; Lisboa, P.J.; Gregson, W. Match Physical Performance of Elite Female Soccer Players during International Competition. *J. Strength Cond. Res.* **2017**, *31*, 2379–2387. [CrossRef]
21. Andersson, H.Å.; Randers, M.B.; Heiner-Møller, A.; Krustrup, P.; Mohr, M. Elite Female Soccer Players Perform More High-Intensity Running When Playing in International Games Compared with Domestic League Games. *J. Strength Cond. Res.* **2010**, *24*, 912–919. [CrossRef]
22. Fernandes, R.; Brito, J.P.; Vieira, L.H.P.; Martins, A.D.; Clemente, F.M.; Nobari, H.; Reis, V.M.; Oliveira, R. In-Season Internal Load and Wellness Variations in Professional Women Soccer Players: Comparisons between Playing Positions and Status. *Int. J. Environ. Res. Public Health* **2021**, *18*, 12817. [CrossRef]
23. Curtis, R.M.; Huggins, R.A.; Benjamin, C.L.; Sekiguchi, Y.; Arent, S.M.; Armwald, B.C.; Pullara, J.M.; West, C.A.; Casa, D.J. Seasonal Accumulated Workloads in Collegiate Men's Soccer: A Comparison of Starters and Reserves. *J. Strength Cond. Res.* **2021**, *35*, 3184–3189. [CrossRef]
24. Giménez, J.V.; Leicht, A.S.; Gomez, M.A. Physical Performance Differences between Starter and Non-Starter Players during Professional Soccer Friendly Matches. *J. Hum. Kinet.* **2019**, *69*, 283–291. [CrossRef]
25. Hills, S.P.; Barrett, S.; Thoseby, B.; Kilduff, L.P.; Barwood, M.J.; Radcliffe, J.N.; Cooke, C.B.; Russell, M. Quantifying the Peak Physical Match-Play Demands of Professional Soccer Substitutes following Pitch-Entry: Assessing Contextual Influences. *Res. Q. Exerc. Sport* **2022**, *93*, 270–281. [CrossRef]
26. Gai, Y.; Volossovitch, A.; Leicht, A.S.; Gómez, M.-Á. Technical and Physical Performances of Chinese Super League Soccer Players Differ According to Their Playing Status and Position. *Int. J. Perform. Anal. Sport* **2019**, *19*, 878–892. [CrossRef]
27. Trewin, J.; Meylan, C.; Varley, M.C.; Cronin, J. The Match-to-Match Variation of Match-Running in Elite Female Soccer. *J. Sci. Med. Sport* **2018**, *21*, 196–201. [CrossRef] [PubMed]
28. Trewin, J.; Meylan, C.; Varley, M.C.; Cronin, J.; Ling, D. Effect of Match Factors on the Running Performance of Elite Female Soccer Players. *J. Strength Cond. Res.* **2018**, *32*, 2002–2009. [CrossRef]
29. Mohr, M.; Krustrup, P.; Andersson, H.; Kirkendal, D.; Bangsbo, J. Match Activities of Elite Women Soccer Players at Different Performance Levels. *J. Strength Cond. Res.* **2008**, *22*, 341–349. [CrossRef]
30. Datson, N.; Hulton, A.; Andersson, H.; Lewis, T.; Weston, M.; Drust, B.; Gregson, W. Applied Physiology of Female Soccer: An Update. *Sports Med.* **2014**, *44*, 1225–1240. [CrossRef]
31. Vescovi, J.D.; Favero, T.G. Motion Characteristics of Women's College Soccer Matches: Female Athletes in Motion (FAiM) Study. *Int. J. Sports Physiol. Perform.* **2014**, *9*, 405–414. [CrossRef] [PubMed]
32. Risso, F.G.; Jalilvand, F.; Orjalo, A.J.; Moreno, M.R.; Davis, D.L.; Birmingham-Babauta, S.A.; Stokes, J.J.; Stage, A.A.; Liu, T.M.; Giuliano, D.V. Physiological Characteristics of Projected Starters and Non-Starters in the Field Positions from a Division I Women's Soccer Team. *Int. J. Exerc. Sci.* **2017**, *10*, 568.
33. McLean, B.D.; Petrucelli, C.; Coyle, E.F. Maximal Power Output and Perceptual Fatigue Responses during a Division I Female Collegiate Soccer Season. *J. Strength Cond. Res.* **2012**, *26*, 3189–3196. [CrossRef]
34. Lorenzo-Martínez, M.; Padrón-Cabo, A.; Rey, E.; Memmert, D. Analysis of Physical and Technical Performance of Substitute Players in Professional Soccer. *Res. Q. Exerc. Sport* **2021**, *92*, 599–606. [CrossRef] [PubMed]
35. Jagim, A.R.; Murphy, J.; Schaefer, A.Q.; Askow, A.T.; Luedke, J.A.; Erickson, J.L.; Jones, M.T. Match Demands of Women's Collegiate Soccer. *Sports* **2020**, *8*, 87. [CrossRef] [PubMed]
36. Akyildiz, Z.; Yildiz, M.; Clemente, F.M. The Reliability and Accuracy of Polar Team Pro GPS Units. *Proc. Inst. Mech. Eng. Part P J. Sports Eng. Technol.* **2020**, *236*, 83–89. [CrossRef]
37. Jennings, D.; Cormack, S.; Coutts, A.J.; Boyd, L.J.; Aughey, R.J. Variability of GPS Units for Measuring Distance in Team Sport Movements. *Int. J. Sports Physiol. Perform.* **2010**, *5*, 565–569. [CrossRef] [PubMed]
38. Hopkins, W.; Marshall, S.; Batterham, A.; Hanin, J. Progressive Statistics for Studies in Sports Medicine and Exercise Science. *Med. Sci. Sports Exerc.* **2009**, *41*, 3. [CrossRef] [PubMed]
39. Padrón-Cabo, A.; Rey, E.; Vidal, B.; García-Nuñez, J. Work-Rate Analysis of Substitute Players in Professional Soccer: Analysis of Seasonal Variations. *J. Hum. Kinet.* **2018**, *65*, 165–174. [CrossRef]
40. Bradley, P.S.; Noakes, T.D. Match Running Performance Fluctuations in Elite Soccer: Indicative of Fatigue, Pacing or Situational Influences? *J. Sports Sci.* **2013**, *31*, 1627–1638. [CrossRef]
41. Gabbett, T.J. The Training—Injury Prevention Paradox: Should Athletes Be Training Smarter and Harder? *Br. J. Sports Med.* **2016**, *50*, 273–280. [CrossRef]

42. Rogalski, B.; Dawson, B.; Heasman, J.; Gabbett, T.J. Training and Game Loads and Injury Risk in Elite Australian Footballers. *J. Sci. Med. Sport* **2013**, *16*, 499–503. [CrossRef]
43. Dalen, T.; Jørgen, I.; Gertjan, E.; Havard, H.G.; Ulrik, W. Player Load, Acceleration, and Deceleration during Forty-Five Competitive Matches of Elite Soccer. *J. Strength Cond. Res.* **2016**, *30*, 351–359. [CrossRef]
44. Osgnach, C.; Poser, S.; Bernardini, R.; Rinaldo, R.; Di Prampero, P.E. Energy Cost and Metabolic Power in Elite Soccer: A New Match Analysis Approach. *Med. Sci. Sports Exerc.* **2010**, *42*, 170–178. [CrossRef] [PubMed]
45. Dalen, T.; Sandmæl, S.; Stevens, T.G.; Hjelde, G.H.; Kjøsnes, T.N.; Wisløff, U. Differences in Acceleration and High-Intensity Activities between Small-Sided Games and Peak Periods of Official Matches in Elite Soccer Players. *J. Strength Cond. Res.* **2021**, *35*, 2018–2024. [CrossRef]
46. Casamichana, D.; Martín-García, A.; Díaz, A.G.; Bradley, P.S.; Castellano, J. Accumulative Weekly Load in a Professional Football Team: With Special Reference to Match Playing Time and Game Position. *Biol. Sport* **2021**, *39*, 115–124. [CrossRef]
47. Anderson, L.; Orme, P.; Di Michele, R.; Close, G.L.; Milsom, J.; Morgans, R.; Drust, B.; Morton, J.P. Quantification of Seasonal-Long Physical Load in Soccer Players with Different Starting Status from the English Premier League: Implications for Maintaining Squad Physical Fitness. *Int. J. Sports Physiol. Perform.* **2016**, *11*, 1038–1046. [CrossRef]
48. Fields, J.; Merrigan, J.; Feit, M.K.; Jones, M. Practice Versus Game External Load Measures in Starters and Non-Starters of a Men's Collegiate Soccer Team. *Int. J. Strength Cond.* **2021**, *1*, 1–8.
49. Nobari, H.; Oliveira, R.; Siahkouhian, M.; Pérez-Gómez, J.; Cazan, F.; Ardigò, L.P. Variations of Accelerometer and Metabolic Power Global Positioning System Variables across a Soccer Season: A within-Group Study for Starters and Non-Starters. *Appl. Sci.* **2021**, *11*, 6747. [CrossRef]
50. Dalen, T.; Lorås, H. Monitoring Training and Match Physical Load in Junior Soccer Players: Starters versus Substitutes. *Sports* **2019**, *7*, 70. [CrossRef] [PubMed]
51. Barreira, J.; Nakamura, F.Y.; Ferreira, R.; Pereira, J.; Aquino, R.; Figueiredo, P. Season Match Loads of a Portuguese Under-23 Soccer Team: Differences between Different Starting Statuses throughout the Season and Specific Periods within the Season Using Global Positioning Systems. *Sensors* **2022**, *22*, 6379. [CrossRef]
52. Hernández, D.; Sánchez, M.; Martin, V.; Benéitez-Andrés, E.; Sánchez-Sánchez, J. Contextual Variables and Weekly External Load in a Semi-Professional Football Team. *Apunt. Educ. Física Esports* **2021**, *146*, 61–67.
53. Reche-Soto, P.; Rojas-Valverde, D.; Bastida-Castillo, A.; Gómez-Carmona, C.D.; Rico-González, M.; Palucci Vieira, L.H.; Paolo Ardigò, L.; Pino-Ortega, J. Using Ultra-Wide Band to Analyze Soccer Performance through Load Indicators during a Full Season: A Comparison between Starters and Non-Starters. *Appl. Sci.* **2022**, *12*, 12675. [CrossRef]
54. Castillo-Rodríguez, A.; González-Téllez, J.L.; Figueiredo, A.; Chinchilla-Minguet, J.L.; Onetti-Onetti, W. Starters and Non-Starters Soccer Players in Competition: Is Physical Performance Increased by the Substitutions? *BMC Sports Sci. Med. Rehabil.* **2023**, *15*, 33. [CrossRef] [PubMed]

Journal of
Functional Morphology and Kinesiology

MDPI

Article

Countermovement Jump Performance Is Related to Ankle Flexibility and Knee Extensors Torque in Female Adolescent Volleyball Athletes

Vassilios Panoutsakopoulos [1] and Eleni Bassa [2,*]

[1] Biomechanics Laboratory, School of Physical Education and Sport Science at Thessaloniki, Aristotle University of Thessaloniki, 54124 Thessaloniki, Greece; bpanouts@phed.auth.gr
[2] Laboratory of Evaluation of Human Biological Performance, School of Physical Education and Sport Science at Thessaloniki, Aristotle University of Thessaloniki, 54124 Thessaloniki, Greece
* Correspondence: lbassa@phed.auth.gr; Tel.: +30-23-1099-2246

Abstract: Ankle flexibility and isokinetic knee torque/power generating capacity were previously suggested to contribute or to be correlated to the vertical countermovement jump (CMJ) performance. The aim of this study was to investigate the effect of the passive ankle joint dorsi flexion (θ_{PDF}) and the knee muscle's isokinetic torque and power on the CMJ in adolescent female volleyball players. The θ_{PDF} at a knee extension angle of 140 degrees were measured for 37 female post-pubertal volleyball players. Then, the players were assigned to either the flexible (n = 10) or inflexible (n = 14) groups according to earlier recommended criteria. Testing included the CMJ with and without an arm swing, and maximal knee extensions and flexions in 3 angular velocities on an isokinetic dynamometer. CMJ height performed with or without an arm swing ($r_{(22)}$ = 0.563, p = 0.040 and $r_{(22)}$ = 0.518, p = 0.009, respectively) and relative power ($r_{(22)}$ = 0.517, p = 0.010 and $r_{(22)}$ = 0.446, p = 0.030, respectively) were positively correlated with the extensors' torque at 180°/s and were negatively correlated with the flexibility level of the dominant side ankle ($r_{(22)}$ = −0.529, p = 0.008 and $r_{(22)}$ = −0.576, p = 0.030, respectively). A moderate positive correlation was also revealed between the CMJ height with and without an arm swing and the power of the non-dominant knee extensors ($r_{(22)}$ = 0.458, p = 0.024 and $r_{(22)}$ = 0.402, p = 0.049, respectively) and flexors ($r_{(22)}$ = 0.484, p = 0.016 and $r_{(22)}$ = 0.477, p = 0.018, respectively). Results of the 2 × 2 repeated ANOVA measurements revealed that flexible players jumped significantly ($p < 0.05$) higher during the CMJs, whilst there was a group effect only on the isokinetic knee extensor muscles' torque. In conclusion, a more flexible ankle joint and a higher isokinetic knee extensor's torque generating capacity resulted in higher CMJ performance. Therefore, ankle flexibility should be emphasized in training and is suggested to be included in preseason screening tests of youth female volleyball players.

Keywords: biomechanics; sports performance; vertical jumps; isokinetics; stretch–shortening cycle; range of motion; ankle dorsiflexion; power; laterality; pubescent

Citation: Panoutsakopoulos, V.; Bassa, E. Countermovement Jump Performance Is Related to Ankle Flexibility and Knee Extensors Torque in Female Adolescent Volleyball Athletes. *J. Funct. Morphol. Kinesiol.* **2023**, *8*, 76. https://doi.org/10.3390/jfmk8020076

Academic Editor: Diego Minciacchi

Received: 13 May 2023
Revised: 1 June 2023
Accepted: 5 June 2023
Published: 7 June 2023

1. Introduction

Volleyball is a team sport, characterized by intermittent periods of high-intensity activities, interspersed by recovery periods of low intensity activities [1]. An increased jumping ability is considered a crucial fitness component for volleyball high level performance as point-scoring actions are mainly jump-based [2]. In addition, the vertical jump ability is highly associated with the attack action's success [3].

A volleyball player's jumping performance is the most assessed through squat jump (SJ), countermovement (CMJ), or/and drop jump (DJ) [4,5]. The SJ evaluates the jumping ability with only a concentric muscle action, whilst CMJ and DJ involve the utilization of the stretch–shortening cycle [6]. During vertical jumps, volleyball players mainly use a full arm swing with the arms initially swinging backward and then moving forward [7]. A coordinated arm swing shortens the braking phase and prolongs the accelerating phase of the jump [7], resulting in an enhanced jumping height and power output [8].

Due to the specificity of the volleyball game demands, it is imperative for players to possess not only coordination but also appropriate levels of strength and power [9]. A previous simulation training study using a musculoskeletal model of the major muscle groups contributing to jumping performance suggested the knee extensors' training as the most effective means for improving it [10], whilst vastii muscles were proven to be great contributors to vertical center of mass (COM) acceleration during a CMJ [11]. Furthermore, high associations of muscular knee isokinetic peak torque and power with a vertical jump's power and height were previously reported, although the kinematics of these two assessments (open vs. closed-chain activity) differ [12]. However, isokinetic knee testing is a common athlete's evaluation that enables the determination of asymmetries (a) between the dominant and non-dominant limb—via the calculation of inter-limb torque deficit [13]—and (b) between knee flexors and extensors muscles via the conventional ratio, which, according to recent review [14], should be performed during both the preseason and in-season period for the better screening of deficits through a training macrocycle.

Performing a vertical jump requires mechanical energy generated by the proximal muscles to be transferred to a distal joint during the impulse. This energy transfer is facilitated by the bi-articular lower extremity muscles' function [15] and flows from the hip to the knee and finally through the ankle joint, which contributes (~23%) via its plantar flexion to the take-off velocity [16]. The level of contribution of the ankle joint is dependent upon the torque generating capacity of the ankle plantar flexors with the bi-articular gastrocnemius muscle facilitating the energy flow because of the lag in its stimulation onset times [17]. An ankle joint's range of motion and limited passive ankle dorsi flexion, in particular, is considered to be an important factor affecting jumping performance, as more flexible individuals outperform those of poor flexibility in jumping scores [18,19]. It was previously reported that ankle muscles' strength may be determined by the range of motion [20]. Therefore, improving an ankle's range of motion could possibly enhance jumping performance [21] by concurrently diminishing the possibility of injury. This is of importance for volleyball players, as ankle sprains are among the most common injuries they experience [22].

Reduced ankle mobility previously resulted in an impaired jumping performance in SJ in female adolescent volleyball players [23], suggesting the ankle range of motion as a crucial mediator of concentric-only jumping performance in the certain sport and age-group of female athletes. However, CMJ may be considered a more functional test for assessing vertical jumping performance in volleyball players, as it involves the use of the stretch–shortening cycle (SSC) and is highly associated with the spike jump performed during a volleyball match in both attacking- and serving-jump actions [3]. Despite the fact that inter-limb asymmetry is not evident in the bilateral CMJ [24], asymmetry in a single-leg CMJ is negatively related with performance in jumping and sprinting tests in youth team-sport athletes [25,26], with single-leg vertical jumps presenting larger asymmetries compared to horizontal jumping [27]. In line with these findings, volleyball players were found to exhibit a 13.6% greater single leg CMJ performance for the dominant leg [28]. This fact can be attributed to the different mechanical loading in each leg during spike jumps, which may alter the ankle range of motion as well.

Taking into consideration that vertical jumping [29] and isokinetic knee torque [30,31] evaluations are among the most common routine strength and conditioning assessments for volleyball players, it would be of great interest to examine the effect of ankle flexibility of both dominant and non-dominant leg, and knee torque generating capacity on CMJ performance in female adolescent volleyball athletes. We hypothesized that CMJ performance would be positively influenced by both of these previously mentioned variables.

2. Materials and Methods

2.1. Design of the Study

To fulfill the purpose of the study, after the assessment of the flexibility of the ankle joint, measures of the knee extensors and flexors isokinetic torque, and the examination of the biomechanical parameters of the CMJ with and without an arm swing were performed in random order.

2.2. Participants

Thirty-seven ($n = 37$) female pubescent volleyball players (16.5 ± 1.2 yrs, 1.80 ± 0.05 m, 68.5 ± 6.6 kg), selected to join the youth national teams, participated in this study. All participants were pubescent according to the Tanner [32] stages (Tanner stage V). This was also confirmed calculating the maturity offset [33], which was 4.19 ± 0.76 yrs for the inflexible (NFG) and 5.11 ± 0.84 yrs for flexible (FLX) group. Thus, all athletes were characterized as post-PHV. The players participated systematically in their training program (10–12 h/wk), had no injury for a 6-month period prior their evaluation and they were tested at least 24 h after the last strenuous training session. Their laboratory evaluation was a part of a wider physical conditioning screening program. The assessments were conducted in accordance with the Declaration of Helsinki and the Research Ethics Code of the Aristotle University of Thessaloniki.

2.3. Experimental Procedure

Firstly, the anthropometric characteristics of the participants were assessed. Body mass was measured to the nearest 0.1 kg using a digital weight scale (BC-545N, Tanita, Tokyo, Japan). A wall-mounted stadiometer (HR001; Tanita Tokyo, Japan) was used to assess the barefoot standing height to the nearest 0.1 cm. The dominant side was defined based on the preferred striking arm during the volleyball spike [34].

2.3.1. Flexibility Assessment

Before the warm-up, the ankle joint flexibility test was conducted in a random order concerning the ipsilateral (DM) and contralateral (NDM) ankle joint of the dominant side. The passive non-weightbearing ankle joint dorsi flexion (θ_{PDF}) [35], when the knee joint was fully extended (180° = full extension) and at a 140° angle, was measured. However, only the θ_{PDF} scores assessed at a 140° knee angle were further used, as this angle is suggested to consist of the representative lower limb configuration that is similar to the knee angle for vertical jumping execution in the majority of sports [36], and in volleyball sport-specific jumps, in particular [37].

Following this recommendation, θ_{PDF} was measured using a video analysis method [23]. A Panasonic NV-MS4E (Matsushita Electric Industrial Company, Osaka, Japan) camera (sampling frequency: 25 fps) was placed on a tripod (height: 1.2 m) at a distance of 4 m perpendicular to an examination bed. Before the measurement, the recorded field of view was calibrated using a 1.25 m × 1.25 m calibration frame with 10 reference markers. The participants sat and were fixed barefoot at the edge of the examination bed, with a hip angle of 120° [18], and the knee joint being at the edge of the bed at a 140° angle. Custom markers (diameter: 0.01 m) were attached on the tuberosity of the 5th metatarsal, the lateral malleolus, the posterior aspect of the calcaneus, the lateral epicondyle of the femur, and the greater trochanter.

For the measurement of θ_{PDF}, force was applied from an experienced examiner on the plantar surface of the foot to dorsi-flex the ankle joint until a feeling of discomfort was stated by the participants [18]. Afterwards, the captured dorsi-flexion was projected on a COMPLOT 7000 digitizer (Mayline Company Inc., Sheboygan, WI, USA) after attaching the camera with a Citizen 30PC-1EB 1EA projector (Japan CBM Corp., Tokyo, Japan). The extracted coordinates of the digitized anatomical points in a two-dimensional Cartesian coordinate system were used to confirm the 140° knee joint angle and to compute θ_{PDF} with a 2D-DLT analysis method provided by the ANGLES 2004 software (©: Iraklis A. Kollias, Biomechanics Laboratory, Aristotle University of Thessaloniki, Thessaloniki, Greece).

The outcome of the θ_{PDF} measurement led to the formation of 2 experimental groups: the FLX (n = 10, 17.1 ± 0.9 yrs, 1.80 ± 0.04 m, 68.8 ± 5.7 kg), and the NFG (n = 14, 15.7 ± 0.8 yrs, 1.79 ± 0.05 m, 69.0 ± 7.3 kg) group. The cut-off thresholds for FLX and NFG were $\theta_{PDF} < 61°$ and $\theta_{PDF} > 69°$, respectively. These cut-offs are suggested [38] to classify individuals as FLX (7.5th percentile) or NFG (92.5th percentile), respectively, based on the frequency distribution analysis of the Laboratory's database that is comprised of a large cohort of female athletes and physical education students (n > 400). Inclusion in either group was considered only if the inter-limb difference for θ_{PDF} was less than 10°.

2.3.2. Warm Up

A warm-up session followed the measurement of θ_{PDF}. The players cycled for 8 min on an 817E Monark Exercise Cycle (Monark-Crescent AB, Varberg, Sweden). Then, they executed dynamic stretches with a progressively increasing range of motion. Finally, six CMJ, both without (CMJA) and with (CMJF) an arm swing and with increasing intensity from sub-maximum to maximum, were allowed for familiarization with the testing procedure.

2.3.3. Vertical Jumps

Both CMJA and CMJF were performed on an AMTI OR6-5-1 force-plate (AMTI, Newton, MA, USA; sampling frequency: 1 kHz). All participants performed, in a random order, three CMJAs with arms kept akimbo and three CMJFs. The command was to "jump as fast and as high as possible". No specific instruction was provided concerning the knee flexion during the countermovement. The intra-jump interval was 60 s, and the inter-test rest was 3 min.

Force-plate data acquisition and analysis was conducted using the routines of the K-Dynami 2018 (©: Iraklis A. Kollias, Biomechanics Laboratory, Aristotle University of Thessaloniki, Thessaloniki, Greece) software. The recorded vertical ground reaction force (vGRF) data was smoothed with a 2nd-order digital low pass Butterworth recursive filter. The cut-off frequency was set using the sum of residuals method [39] to 20 Hz. The jump height (H_{CMJ}) was calculated from the COM vertical take-off velocity (V_0) that was calculated as the first-time integral of the net vGRF using the trapezoid rule. The spatio-temporal (downward vertical COM displacement-S_D; upward vertical COM displacement-S_U; impulse time-t_C; duration of the propulsion phase-t_{PROP}) and the kinetic (net vGRF-Fz; rate of force development-RFD; peak power-P_{MAX}) parameters of the CMJ tests were extracted based on the vGRF-time series, the participants' mass, and classical equations of motion, as described in detail elsewhere [8]. The arm swing gain was estimated as the percentage chance of H_{CMJ} in CMJF compared to CMJA. The reactive strength index (RSI) was calculated as H_{CMJ}/t_C [40], which is suggested to be an appropriate performance indicator for volleyball players [41]. For further analysis, only the best attempt, as defined by H_{CMJ}, was selected.

2.3.4. Isokinetic Evaluation

Participants performed concentric contractions of knee extensors and flexors seated (hip angle: 115°) on an isokinetic dynamometer (Cybex Norm, CYBEX Division of Lumex, Ronkonkoma, NY, USA). The trunk, waist, and upper thigh were stabilized on the chair using velcro straps to avoid any movement that could impact the measurement quality. Each participant raised the leg in parallel to the ground, correction of gravity was applied, full knee joint extension was checked, and the most prominent point of the medial femoral epicondyle was aligned with the axis of rotation of the dynamometer. Prior to the isokinetic test, participants performed 5 submaximal concentric knee flexions and extensions as familiarization. The isokinetic evaluation included 3 maximal knee extension and flexion trials at the concentric angular velocities of 60°/s and 180°/s performed in a randomized order. Knee range of motion was limited for all subjects from 0° to 90° of their knee flexion.

The participants watched their torque scores on the screen of the dynamometer in order to outperform each previous trial (visual feedback) and were encouraged to perform their best in both movement directions. Inter-set rest was 3 min to avoid any fatigue effects. The highest peak torque and power values assessed at each angular velocity for both knee extensors and flexors were used for further analysis.

2.4. Statistical Analysis

According to the calculations using the G*power software (G*power, v.3.1.9.6, ©Franz Faul, University of Kiel, Kiel, Germany), the final sample size of 24 athletes used was the sample required for the present experimental design and it corresponded to 0.7 power for a 0.22 effect size at $a = 0.05$. The sample size was calculated based on the results of Panoutsakopoulos et al. [23].

All examined parameters were presented as mean ± standard deviation. The Shapiro–Wilk test ($p > 0.05$) and the Levene test ($p > 0.05$) were used to check the normality of distribution and the equality of variance, respectively. A 2 (flexibility; FLX, NFG) × 2 (arm swing: CMJA, CMJF) repeated-measures ANOVA with the Bonferroni adjustment was carried out to compare the main effects of flexibility and arm swing, and the interaction effect between flexibility and arm swing on the kinetic and temporal parameters of the CMJ. A 2 (angular velocity; 60°/s, 180°/s) × 2 (groups: FLX, NFG) repeated-measures ANOVA with the Bonferroni adjustment was carried out to compare the main effect of the angular velocity and group and the interaction effect between angular velocity × group on the torque and power of knee extensors and flexors and the conventional ratio of DM and NDM limbs. An Independent Samples t-test was run to check possible group differences in the CMJ gain due to the arm swing. Hedges' g was used to interpret the effect size of the comparison. A Pearson correlation coefficient (r) was computed to assess the linear relationship between the H_{CMJ} in CMJA and CMJF and the knee flexors' and extensors' torque and power.

The statistical analyses were conducted with the IBM SPSS Statistics v.27.0.1.0 software (International Business Machines Corp., Armonk, NY, USA). The level of significance was set at $a = 0.05$ for all analyses.

3. Results

3.1. Passive Ankle Dorsi Flexion

The results of the flexibility measurements are depicted in Figure 1. A significant effect of laterality on θ_{PDF} ($F_{(1,22)} = 38.89$, $p < 0.001$, $\eta_p^2 = 0.64$), a significant group effect (FLX > NFG; $F_{(1,22)} = 57.01$, $p < 0.001$, $\eta_p^2 = 0.72$), and an interaction of laterality × group ($F_{(1,22)} = 34.49$, $p < 0.001$, $\eta_p^2 = 0.61$) was found. The FLX players presented no significant inter-limb difference ($p > 0.05$), whilst the NFG had a lower θ_{PDF} in the DM leg ($p < 0.001$).

3.2. Countermovement Jumps

3.2.1. CMJ Height

A significant effect of the arm swing on H_{CMJ} was found ($F_{(1,22)} = 187.52$, $p < 0.001$, $\eta_p^2 = 0.90$). A significant difference between groups in H_{CMJ} in both conditions (no arm swing and arm swing) was detected, with FLX players presenting higher scores ($F_{(1,22)} = 17.23$, $p < 0.001$, $\eta_p^2 = 0.44$). No significant ($p > 0.05$) interaction of the arm swing × group was found (Table 1).

3.2.2. CMJ Arm Swing Gain

No significant difference was observed for the gain in CMJ due to the arm swing ($t_{(1,23)} = 0.20$, $p = 0.843$, $g = 0.09$). It was 17.7 ± 5.0% for the FLX and 18.3 ± 8.2% for the NFG.

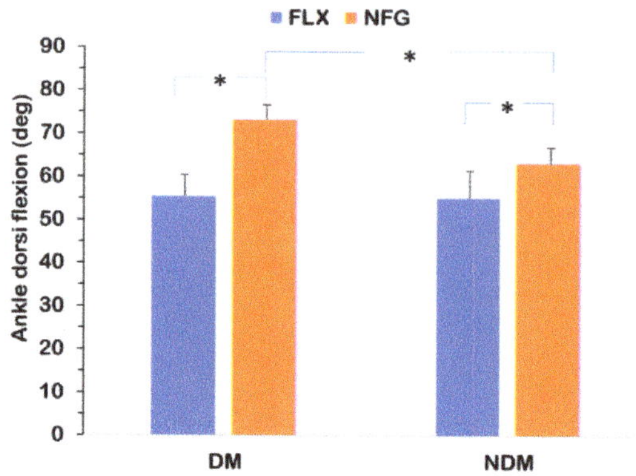

Figure 1. Results of the passive ankle dorsi flexion measurement (FLX: flexible group, n = 10; NFG: inflexible group, n = 14; DM: ipsilateral ankle joint of the preferred arm for the spike; NDM: contralateral ankle joint of the preferred arm for the spike; *: $p < 0.05$).

Table 1. Parameters for the countermovement jump without (CMJA) and with (CMJF) the use of an arm swing in the flexible (FLX, n = 10) and non-flexible (NFG, n = 14) groups.

Parameter	Test	FLX (n = 10) (Mean ± SD)	NFG (n = 14) (Mean ± SD)	Flexibility		Arm Swing		Interaction	
				p	η_p^2	p	η_p^2	p	η_p^2
H_{CMJ} (cm)	CMJA	25.22 ± 3.25	20.21 ± 2.77 *	<0.001	0.44	<0.001	0.90	0.13	0.10
	CMJF	29.74 ± 4.29 #	23.81 ± 2.89 *#						
V_0 (m/s)	CMJA	2.22 ± 0.15	2.00 ± 0.14 *	<0.001	0.42	<0.001	0.90	0.43	0.03
	CMJF	2.41 ± 0.19 #	2.16 ± 0.13 *#						
FZmax (N/kg)	CMJA	2.43 ± 0.21	2.21 ± 0.25 *	0.01	0.26	0.48	0.02	0.38	0.04
	CMJF	2.42 ± 0.18	2.31 ± 0.16						
RFDmax (kN/s)	CMJA	10.40 ± 4.10	8.43 ± 4.22	0.72	0.01	0.14	0.10	0.90	0.12
	CMJF	7.55 ± 2.01 #	8.64 ± 3.55						
P_{MAX} (W/kg)	CMJA	24.63 ± 2.83	19.31 ± 2.75 *	<0.001	0.47	<0.001	0.80	0.90	0.001
	CMJF	31.22 ± 4.74 #	25.72 ± 3.20 *#						
S_D (cm)	CMJA	−30.01 ± 5.12	−31.01 ± 3.79	0.96	0.00	0.18	0.08	0.37	0.04
	CMJF	−29.95 ± 4.11	−29.12 ± 4.83						
S_U (cm)	CMJA	50.91 ± 6.05	48.94 ± 4.37	0.07	0.14	<0.001	0.57	0.58	0.02
	CMJF	53.87 ± 4.25	49.52 ± 5.49						
t_C (ms)	CMJA	597.60 ± 126.74	638.36 ± 101.55	0.59	0.01	0.50	0.02	0.57	0.02
	CMJF	651.50 ± 141.72	642.64 ± 131.12						
t_{PROP} (ms)	CMJA	306.20 ± 36.19	332.00 ± 37.66	0.23	0.06	0.23	0.07	0.43	0.03
	CMJF	334.70 ± 56.92	338.21 ± 45.85						
RSI (m/s)	CMJA	0.83 ± 0.12	0.62 ± 0.11	<0.001	0.46	0.017	0.23	0.92	0.002
	CMJF	0.92 ± 0.22	0.72 ± 0.12						

NOTE: H_{CMJ}: jump height; V_0: body center of mass vertical take-off velocity; FZmax: peak net vertical ground reaction force; RFDmax: peak rate of force development; P_{MAX}: peak power; S_D: downward center of mass vertical displacement; S_U: upward center of mass vertical displacement; t_C: duration of the impulse; t_{PROP}: duration of the propulsive phase; RSI: reactive strength index; *: $p < 0.05$ vs. FLX; #: $p < 0.05$ vs. CMJA.

3.2.3. CMJ Biomechanics

No significant effect of arm swing on the CMJ peak net vGRF relative to body mass was found ($p > 0.05$) but a significant group effect was detected ($F_{(1,22)} = 7.73$, $p < 0.01$, $\eta_p^2 = 0.26$), since FLX presented higher values than NFG (Table 1). A significant effect of arm swing on the CMJ relative power ($F_{(1,22)} = 88.70$, $p < 0.001$, $\eta_p^2 = 0.80$) and RSI ($F_{(1,22)} = 6.74$, $p = 0.017$, $\eta_p^2 = 0.23$) was found.

A significance between the groups' difference in CMJ relative power and RSI in both conditions (no arm swing and swing) was detected, with the more flexible athletes presenting higher power ($F_{(1,22)} = 19.62$, $p < 0.001$, $\eta_p^2 = 0.47$) and RSI ($F_{(1,22)} = 19.03$,

$p < 0.001$, $\eta_p^2 = 0.46$) values. Concerning the maximum RFD, as well as S_D, S_U, t_C, and t_{PROP}, no significant effect of the arm swing ($p > 0.05$) and no significant group effect ($p > 0.05$) was found. Finally, no significant interaction of the arm swing × group was found for all of the above-mentioned parameters ($p > 0.05$).

3.3. Isokinetic Tests

3.3.1. Isokinetic Torque

An angular velocity effect was found for the knee flexors' torque of the DM ($F_{(1,22)} = 184.64$, $p < 0.001$, $\eta_p^2 = 0.89$) and NDM ($F_{(1,22)} = 306.24$, $p < 0.001$, $\eta_p^2 = 0.93$) lower limbs (Table 2). Neither a significance between the groups' difference in the dominant knee flexors' torque nor an interaction of the angular velocity × group was found ($p > 0.05$).

Table 2. Isokinetic torque values of knee extensors (Ext) and flexors (Flex) at 60°/s and 180°/s for the dominant (DM) and the non-dominant (NDM) leg in the flexible (FLX, $n = 10$) and non-flexible (NFG, $n = 14$) groups.

Laterality	Torque (Nm)	FLX ($n = 10$) (Mean ± SD)	NFG ($n = 14$) (Mean ± SD)	Group		Angular Velocity		Interaction	
				p	η_p^2	p	η_p^2	p	η_p^2
DM	Ext 60°/s Ext 180°/s	186.10 ± 25.77 129.50 ± 20.04 #	163.71 ± 25.03 * 111.50 ± 16.12 *#	0.03	0.20	<0.001	0.95	0.41	0.03
NDM	Ext 60°/s Ext 180°/s	181.20 ± 27.50 123.80 ± 17.43 #	154.21 ± 20.61 * 108.36 ± 15.98 *#	0.01	0.25	<0.001	0.92	0.10	0.01
DM	Flex 60°/s Flex 180°/s	98.90 ± 17.79 60.60 ± 17.49 #	92.50 ± 17.43 57.00 ± 13.49 #	0.43	0.03	<0.001	0.89	0.61	0.01
NDM	Flex 60°/s Flex 180°/s	94.40 ± 21.21 59.00 ± 17.99 #	90.64 ± 16.56 57.00 ± 14.54 #	0.68	0.01	<0.001	0.93	0.66	0.01

NOTE: Ext 60°/s: torque of knee extensors at 60°/s; Ext 180°/s: torque of knee extensors at 180°/s; Flex 60°/s: torque of knee flexors at 60°/s; Flex 180°/s: torque of knee flexors at 180°/s; *: $p < 0.05$ vs. FLX; #: $p < 0.05$ vs. 60°/s.

An angular velocity effect on the knee extensors' torque of the DM ($F_{(1,22)} = 443.69$, $p < 0.001$, $\eta_p^2 = 0.95$) and NDM ($F_{(1,22)} = 244.25$, $p < 0.001$, $\eta_p^2 = 0.92$) leg was found. A significant between the groups' difference in the knee extensors' torque of both lower limbs was found ($F_{(1,22)} = 5.394$, $p < 0.030$, $\eta_p^2 = 0.20$ and $F_{(1,22)} = 7.413$, $p < 0.012$, $\eta_p^2 = 0.25$ for DM and NDM, respectively), with FLX presenting a higher knee extensors' torque. No significant interaction of the angular velocity × group was found ($p > 0.05$).

3.3.2. Inter-Limb Torque Deficit

No effect of the muscle group on inter-limb deficit was found, no difference between the groups was detected, and no interaction between the muscle groups × group difference was found ($p > 0.05$).

3.3.3. Conventional Ratio

An angular velocity effect on the conventional ratio of the DM and NDM leg was found, with the ratio being higher at 60°/s ($F_{(1,22)} = 16.85$, $p < 0.001$, $\eta_p^2 = 0.43$ and $F_{(1,22)} = 16.85$, $p < 0.001$, $\eta_p^2 = 0.43$ for DM and NDM, respectively). No difference between the groups and no interaction of the angular velocity × group was found ($p > 0.05$).

3.3.4. Isokinetic Power

An angular velocity effect on the knee flexors' power of the DM ($F_{(1,22)} = 63.28$, $p < 0.001$, $\eta_p^2 = 0.74$) and NDM ($F_{(1,22)} = 65.52$, $p < 0.001$, $\eta_p^2 = 0.75$) leg was found (Table 3), as knee flexors' power presented higher values at 180°/s compared to 60°/s. In addition, an angular velocity effect on the knee extensors' power of the DM ($F_{(1,22)} = 194.37$, $p < 0.001$, $\eta_p^2 = 0.90$) and the NDM ($F_{(1,22)} = 305.48$, $p < 0.001$, $\eta_p^2 = 0.93$) leg was observed, with the knee extensors' power presenting higher values at 180°/s compared to 60°/s. Neither a significance between the groups' difference in DM and NDM knee flexors' power, nor a significance between the groups' difference in DM and NDM knee extensors' power was found ($p > 0.05$). Finally, no interaction of the angular velocity × group on isokinetic power was revealed ($p > 0.05$).

Table 3. Isokinetic power values of knee extensors (Ext) and flexors (Flex) at 60°/s and 180°/s for the dominant (DM) and the non-dominant (NDM) leg in the flexible (FLX, $n = 10$) and non-flexible (NFG, $n = 14$) groups.

Laterality	Power (W)	FLX ($n = 10$) (Mean ± SD)	NFG ($n = 14$) (Mean ± SD)	Group		Angular Velocity		Interaction	
				p	η_p^2	p	η_p^2	p	η_p^2
DM	Ext 60°/s Ext 180°/s	111.46 ± 27.21 207.55 ± 65.22 #	112.86 ± 19.62 209.64 ± 31.26 #	0.90	0.001	<0.001	0.90	0.96	0.00
NDM	Ext 60°/s Ext 180°/s	117.90 ± 21.39 230.70 ± 46.03 #	112.06 ± 16.64 204.19 ± 34.35 #	0.17	0.08	<0.001	0.93	0.09	0.12
DM	Flex 60°/s Flex 180°/s	66.40 ± 10.90 104.08 ± 31.11 #	64.04 ± 10.97 103.17 ± 26.26 #	0.83	0.002	<0.001	0.74	0.88	0.00
NDM	Flex 60°/s Flex 180°/s	68.31 ± 20.52 1119.63 ± 56.45 #	60.50 ± 10.35 101.91 ± 22.76 #	0.27	0.06	<0.001	0.75	0.40	0.03

NOTE: Ext 60°/s: power of knee extensors at 60°/s; Ext 180°/s: power of knee extensors at 180°/s; Flex 60°/s: power of knee flexors at 60°/s; Flex 180°/s: power of knee flexors at 180°/s; #: $p < 0.05$ vs. 60°/s.

3.4. Correlations

There was a significant positive moderate correlation between the CMJA and CMJF H_{CMJ} and the extensors' torque of the DM leg at 180°/s ($r_{(22)} = 0.563$, $p = 0.040$ and $r_{(22)} = 0.518$, $p = 0.009$, respectively). There was also a significant positive moderate correlation between the CMJA and CMJF H_{CMJ} and the NDM leg extensors' torque at 180°/s ($r_{(22)} = 0.514$, $p = 0.010$ and $r_{(22)} = 0.456$, $p = 0.025$, respectively).

A significant positive moderate correlation between the CMJA and CMJF and the flexors' power of the NDM leg at 180°/s was observed ($r_{(22)} = 0.484$, $p = 0.016$ and $r_{(22)} = 0.477$, $p = 0.018$, respectively). In addition, a significant positive moderate correlation between the CMJA and CMJF and the NDM extensors' power at 180°/s was revealed ($r_{(22)} = 0.458$, $p = 0.024$ and $r_{(22)} = 0.402$, $p = 0.049$, respectively).

A significant moderate correlation between the DM leg extensors' torque at 180°/s and the CMJA and CMJF relative power was detected ($r_{(22)} = 0.517$, $p = 0.010$ and $r_{(22)} = 0.446$, $p = 0.030$, respectively). This was also observed for the NDM leg, since a significant moderate correlation between the NDM leg extensors' torque at 180°/s and the CMJA and CMJF relative power was detected ($r_{(22)} = 0.461$, $p = 0.020$ and $r_{(22)} = 0.414$, $p = 0.040$, respectively).

A significant negative moderate correlation between the CMJA and CMJF and the θ_{PDF} of the DM leg was detected ($r_{(22)} = -0.529$, $p = 0.008$ and $r_{(22)} = -0.576$, $p = 0.030$, respectively). Similarly, a significant negative correlation between the CMJA and CMJF P_{MAX} and the θ_{PDF} of the DM leg was found ($r_{(22)} = -0.535$, $p = 0.007$ and $r_{(22)} = -0.586$, $p = 0.003$, respectively).

No other significant correlations were revealed.

4. Discussion

The current study examined the hypothesis that the countermovement jump performance with and without arm swing in female pubescent volleyball players could be affected by (a) ankle flexibility, (b) by asymmetries in ankle flexibility between ipsilateral and contralateral to spike arm leg, and/or by (c) knee flexors' and extensors' torque. The hypothesis was confirmed as the FLX players, who presented no inter-limb difference in the passive ankle joint dorsi flexion, jumped higher than the less flexible, who presented a relatively restricted passive ankle joint dorsi flexion at the ipsilateral to spike arm leg. Regarding isokinetic knee torque, the NFG produced lower knee extensors' torque than FLX in both 60°/s and 180°/s. Furthermore, the CMJ height and power output in the trials performed with or without an arm swing were positively correlated with the extensors' torque at 180°/s and were negatively correlated with the flexibility level of DM ankle, suggesting that both parameters are significant mediators of CMJ performance.

The FLX jumped higher in the CMJ than NFG, corroborating previous results in SJ, where ankle flexibility affected the SJ performance in female volleyball players [23]. This could be attributed to the larger force and power output [8,42–45]. In the less flexible or reduced ankle mobility conditions, the reduced contribution of the biarticular gastrocnemius to the energy transfer was previously reported to be counterbalanced with the larger mobility of the torso and the hip joint [36], and an augmented knee mechanical output [46]. The present findings revealed that FLX also showed larger knee extensors' torque compared to the NFG. This indicates the poor capacity of the latter to both produce [47] and transfer energy for the jump. This factor was found not to change in female volleyball players during adolescence [48], but is suggested to discriminate between skilled and players of lesser abilities during this age [49]. Taking this into consideration, one would expect that NFG—although they have a flexibility deficit—to have adapted through the great number of sport-specific jumps performed during volleyball training and they would have probably found an optimal way to perform the CMJs. Although kinematic analysis was beyond the aims of this study, the lower jumping height, force, velocity, and power values in NFG athletes suggested that this was probably not the case.

RSI comprises of a temporal normalization of jump height, categorizing jumping activities in slow or fast. FLX presented higher RSI scores than NFG, proving not only their ability to jump higher than NFG, but also the ability to jump faster, which is considered the desirable way to perform jumps in athletic activities. Therefore, we may also assume that the FLX athletes probably store and release more energy, demonstrating superior and more effective utilization of the SSC [50] during CMJ than their inflexible counterparts. RSI is considered a sensitive indicator of efficient neuromuscular function and lower limb explosiveness in female volleyball athletes [51]. Furthermore, RSI was previously highly correlated with higher force, power, velocity, and impulse during jumps [52]—parameters, in which, FLXs were found to have larger values than the NFGs.

As mentioned above, the parameters interpreting explosiveness in vertical jumps, such as power and reactive strength, were augmented with the arm swing and a flexible ankle joint. The examined groups were different in the force output in the no-arm swing CMJ, but not in the CMJ with an arm swing. The arm swing generates mechanical work that is transferred and imposes a greater load to the lower limb muscles, thus leading to a higher capacity to produce energy for the jump [42,53]. Thus, it seems that the inflexible players used the additional work produced by the arm swing to limit the deficiency in energy production due to the limited mobility of the ankle joint. However, the energy transfer from the upper to the lower limbs should be sequentially synchronized throughout the jump [53]. Past research [18] utilizing kinematical analysis of the CMJ revealed differences between flexible and inflexible individuals in the body configuration and the rotational kinematics of the lower limb joints, concluding that inflexible individuals absorb energy during the eccentric phase of the CMJ that is not compensated during the propulsion phase, thus leading to a decreased jumping performance due to limited movement amplitudes [54]. The lack of a kinematical analysis deprives the extraction of solid evidence concerning this mechanism.

Interestingly, there was a significant interaction of laterality and flexibility with only the NFG presenting significant differences in flexibility between the dominant and non-dominant leg. Laterality differences in the flexibility may be attributed to the different sport-specific demands of jumps for the joint kinematics of each leg in volleyball [9,37]. Volleyball athletes perform a lot of spike jumps that load differently the ipsi- and contra-lateral to spike arm leg. Variations in weight distribution, power development, ankle angle at foot planting, and pressure experienced at ipsi- and contra-lateral leg during foot planting, are different neuromuscular stimuli that could result in specific adaptations [34]. Generally speaking, skeletal muscle tissue may remodel its structure adapting to mechanical loading [55]. Thus, higher forces applied during the sport-specific jumps on the contralateral to spike arm leg may have probably resulted in a higher ankle range of motion compared to the ipsilateral leg in the NFG. This assumption could be further supported by the notion that the higher the force accelerating the ankle motion, the greater the ankle range of motion [56]. Previous research in handball players revealed that laterality, based on hand preference, resulted in significant differences in the active ankle joint range of motion at the selected knee angle where the flexibility test in the present study was conducted [57]. Additionally, a recent study applying a stretching protocol to 1 leg for 12 weeks reported increases in the ankle range of motion of the non-trained leg, a finding that led the researchers speculate that volleyball training, per se, probably affects ankle flexibility [58]. However, such kind of adaptations were not apparent in the FLX, suggesting that a flexible joint probably remains at its level of flexibility in both legs, regardless of the inter-limb loading differences. Furthermore, flexibility differences between groups may also be attributed to an age effect, as the NFG players were younger than the FLX (15.66 ± 0.77 yrs vs. 17.05 ± 0.94 yrs, respectively). This finding further supports the hypothesis that flexibility may be altered during pubescence by specific sport training [58], since muscles and joints probably have the potential to adapt, while growing, to meet a particular sport-specific performance requirement [59].

Warm-up before any athletic action is considered essential to optimize performance [60]. In detail, the warm-up procedure elevates temperature, which may decrease the viscosity of the tissues [60], resulting in a lower resistance to stretch and an increased joint range of motion [61]. However, in the present study, no warm-up was performed prior the flexibility measurements, because warm-up was found to have no effect on flexibility compared to the control condition (i.e., no warm-up), whilst any increases in flexibility were observed only after stretching [62]. Finally, the present study design is further supported by similar research in the literature, where warm-up was performed after the flexibility testing session and before the execution of the jumping tests [18].

We are aware that our research may have limitations. The isokinetic evaluation was conducted only in the knee joint as this assessment of the knee extensor and flexor muscles is a common practice in preseason athletes' screening [63]. An additional ankle torque evaluation would probably offer better insight in the probable contribution of ankle muscles in the CMJ performance in female volleyball athletes of different ankle flexibility. Furthermore, a kinematic analysis could provide additional information about the differences between flexible and inflexible young female volleyball players in terms of posture and knee joint rotational kinematics. Future research using kinematic recordings and isokinetic evaluation of ankle joint muscles needs to be conducted. Finally, asymmetry is affected by the training season [27]. The fact that flexibility asymmetry was assessed in a single day during the end-season may have affected not so much the magnitude, but rather the direction of asymmetry [64], probably resulting in different bilateral CMJ scores, compared to scores assessed during other training seasons, a case that would be interesting to be examined as well.

5. Conclusions

Vertical jump performance is crucial for high level performance in volleyball. According to the findings of this study, female adolescent volleyball players who present a high ankle joint flexibility and can exert high isokinetic knee extensors' torque outperform their less flexible and weaker counterparts in the vertical jump performance. Therefore, to improve the vertical jumping ability of female adolescent volleyball athletes throughout their long-term training, coaches are encouraged not only to check their athletes for knee muscles' torque deficits, but also for any possible ankle flexibility deficits or asymmetries through different training seasons, and to train them accordingly in order to augment performance and concurrently limit injury occurrence possibility. As adolescence is an important time for laying the foundations for long-term athletic development, an individual approach, in both flexibility and strength development, is suggested for maximizing training effects whilst supporting sound physical development.

Author Contributions: Conceptualization, E.B. and V.P.; methodology, E.B. and V.P.; validation, E.B. and V.P.; formal analysis, E.B. and V.P.; investigation, E.B. and V.P.; resources, E.B. and V.P.; data curation, E.B. and V.P.; writing—original draft preparation, E.B. and V.P.; writing—review and editing, E.B. and V.P.; visualization, E.B. and V.P.; supervision, E.B.; project administration, V.P. All authors have read and agreed to the published version of the manuscript.

Funding: This research received no external funding.

Institutional Review Board Statement: The study was conducted according to the guidelines of the Declaration of Helsinki and to the Institution's Research Committee Ethics Code. The research was completed under the auspices of the Research Committee of Aristotle University of Thessaloniki (protocol code: 7233/09.11.1997, approved 9 November 1997).

Informed Consent Statement: Informed consent was obtained from all subjects involved in the study.

Data Availability Statement: The data that were used in the present study can be provided by the corresponding author upon reasonable request.

Conflicts of Interest: The authors declare no conflict of interest.

References

1. Calleja-Gonzalez, J.; Mielgo-Ayuso, J.; Sanchez-Ureña, B.; Ostojic, S.M.; Terrados, N. Recovery in volleyball. *J. Sports Med. Phys. Fit.* **2019**, *59*, 982–993. [CrossRef] [PubMed]
2. García-de-Alcaraz, A.; Ramírez-Campillo, R.; Rivera-Rodríguez, M.; Romero-Moraleda, B. Analysis of jump load during a volleyball season in terms of player role. *J. Sci. Med. Sport* **2020**, *23*, 973–978. [CrossRef] [PubMed]
3. Berriel, G.P.; Schons, P.; Costa, R.R.; Oses, V.H.S.; Fischer, G.; Pantoja, P.D.; Kruel, L.F.M.; Peyré-Tartaruga, L.A. Correlations between jump performance in block and attack and the performance in official games, squat jumps, and countermovement jumps of professional volleyball players. *J. Strength Cond. Res.* **2021**, *35* (Suppl. S2), S64–S69. [CrossRef] [PubMed]
4. Silva, A.F.; Clemente, F.M.; Lima, R.; Nikolaidis, P.T.; Rosemann, T.; Knechtle, B. The effect of plyometric training in volleyball players: A systematic review. *Int. J. Environ. Res. Public Health* **2019**, *16*, 2960. [CrossRef]
5. Ziv, G.; Lidor, R. Vertical jump in female and male volleyball players: A review of observational and experimental studies. *Scand J. Med. Sci. Sports* **2010**, *20*, 556–567. [CrossRef]
6. Harman, E. Measurement of human mechanical power. In *Physiological Assessment of Human Fitness*; Maud, P., Foster, C., Eds.; Human Kinetics: Champaign, IL, USA, 2006; pp. 87–113.
7. Vaverka, F.; Jandačka, D.; Zahradník, D.; Uchytil, J.; Farana, R.; Supej, M.; Vodičar, J. Effect of an arm swing on countermovement vertical jump performance in elite volleyball players. *J. Hum. Kinet.* **2016**, *53*, 41–50. [CrossRef]
8. Giatsis, G.; Panoutsakopoulos, V.; Kollias, I.A. Biomechanical differences of arm swing countermovement jumps on sand and rigid surface performed by elite beach volleyball players. *J. Sports Sci.* **2018**, *36*, 997–1008. [CrossRef]
9. Fuchs, P.X.; Fusco, A.; Bell, J.W.; von Duvillard, S.P.; Cortis, C.; Wagner, H. Movement characteristics of volleyball spike jump performance in females. *J. Sci. Med. Sport* **2019**, *22*, 833–837. [CrossRef]
10. Nagano, A.; Gerritsen, K.G. Effects of neuromuscular strength training on vertical jumping performance: A computer simulation study. *J. Appl. Biomech.* **2001**, *17*, 113–128. [CrossRef]
11. Kipp, K.; Kim, H. Relative contributions and capacities of lower extremity muscles to accelerate the body's center of mass during countermovement jumps. *Comput. Methods Biomech. Biomed. Engin.* **2020**, *23*, 914–921. [CrossRef]
12. Yapici, A.; Findikoglu, G.; Dundar, U. Do isokinetic angular velocity and contraction types affect the predictors of different anaerobic power tests? *J. Sports Med. Phys. Fit.* **2016**, *56*, 383–391.
13. Lockie, R.G.; Schultz, A.B.; Jeffriess, M.D.; Callaghan, S.J. The relationship between bilateral differences of knee flexor and extensor isokinetic strength and multi-directional speed. *Isokinet. Exer. Sci.* **2012**, *20*, 211–219. [CrossRef]
14. Kellis, E.; Sahinis, C.; Baltzopoulos, V. Is hamstrings-to-quadriceps torque ratio useful for predicting anterior cruciate ligament and hamstring injuries? A systematic and critical review. *J. Sport Health Sci.* **2022**, *19*, 343–358. [CrossRef] [PubMed]
15. Prilutsky, B.I.; Zatsiorsky, V.M. Tendon action of two-joint muscles: Transfer of mechanical energy between joints during jumping, landing, and running. *J. Biomech.* **1994**, *27*, 25–34. [CrossRef] [PubMed]
16. Hubley, C.L.; Wells, R.P. A work energy approach to determine individual joint contributions to vertical jump performance. *Eur. J. Appl. Physiol. Occup. Physiol.* **1983**, *50*, 247–254. [CrossRef]
17. Bobbert, M.F.; van Zandwijk, J.P. Sensitivity of vertical jumping performance to changes in muscle stimulation onset times: A simulation study. *Biol. Cybern.* **1999**, *81*, 101–108. [CrossRef]
18. Papaiakovou, G. Kinematic and kinetic differences in the execution of vertical jumps between people with good and poor ankle joint dorsiflexion. *J. Sports Sci.* **2013**, *31*, 1789–1796. [CrossRef]
19. Papaiakovou, G.; Nikodelis, T.; Panoutsakopoulos, V.; Kollias, I. Effects of initial posture upon vertical squat jump dynamic and kinematic characteristics of subjects with limited ankle joint dorsi flexion. *J. Hum. Mov. Stud.* **2003**, *44*, 311–322.

20. Guillén-Rogel, P.; San Emeterio, C.; Marín, P.J. Associations between ankle dorsiflexion range of motion and foot and ankle strength in young adults. *J. Phys. Ther. Sci.* **2017**, *29*, 1363–1367. [CrossRef]
21. Yun, S.J.; Kim, M.H.; Weon, J.H.; Kim, Y.; Jung, S.H.; Kwon, O.Y. Correlation between toe flexor strength and ankle dorsiflexion ROM during the countermovement jump. *J. Phys. Ther. Sci.* **2016**, *28*, 2241–2244. [CrossRef]
22. Chandran, A.; Morris, S.N.; Lempke, L.B.; Boltz, A.J.; Robison, H.J.; Collins, C.L. Epidemiology of injuries in national collegiate athletic association women's volleyball: 2014–2015 through 2018–2019. *J. Athl. Train.* **2021**, *56*, 666–673. [CrossRef] [PubMed]
23. Panoutsakopoulos, V.; Kotzamanidou, M.C.; Papaiakovou, G.; Kollias, I.A. The ankle joint range of motion and its effect on squat jump performance with and without arm swing in adolescent female volleyball players. *J. Funct. Morphol. Kinesiol.* **2021**, *6*, 14. [CrossRef] [PubMed]
24. Heishman, A.; Daub, B.; Miller, R.; Brown, B.; Freitas, E.; Bemben, M. Countermovement jump inter-limb asymmetries in collegiate basketball players. *Sports* **2019**, *7*, 103. [CrossRef]
25. Arboix-Alió, J.; Buscà, B.; Busquets, A.; Aguilera-Castells, J.; de Pablo, B.; Montalvo, A.M.; Fort-Vanmeerhaeghe, A. Relationship between inter-limb asymmetries and physical performance in rink hockey players. *Symmetry* **2020**, *12*, 2035. [CrossRef]
26. Bishop, C.; Read, P.; McCubbine, J.; Turner, A. Vertical and horizontal asymmetries are related to slower sprinting and jump performance in elite youth female soccer players. *J. Strength Cond. Res.* **2018**, *35*, 56–63. [CrossRef]
27. Fort-Vanmeerhaeghe, A.; Bishop, C.; Buscà, B.; Vicens-Bordas, J.; Arboix-Alió, J. Seasonal variation of inter-limb jumping asymmetries in youth team-sport athletes. *J. Sports Sci.* **2021**, *39*, 2850–2858. [CrossRef]
28. Bartol, V.; Vauhnik, R.; Rugelj, D. Influence of the sport specific training background on the symmetry of the single legged vertical counter movement jump among female ballet dancers and volleyball players. *Heliyon* **2022**, *8*, e10669. [CrossRef]
29. Sheppard, J.M.; Newton, R.U. Long-term training adaptations in elite male volleyball players. *J. Strength Cond. Res.* **2012**, *26*, 2180–2184. [CrossRef]
30. Sattler, T.; Sekulic, D.; Esco, M.R.; Mahmutovic, I.; Hadzic, V. Analysis of the association between isokinetic knee strength with offensive and defensive jumping capacity in high-level female volleyball athletes. *J. Sci. Med. Sport* **2015**, *18*, 613–618. [CrossRef]
31. Wilkosz, P.; Kabacinski, J.; Mackala, K.; Murawa, M.; Ostarello, J.; Rzepnicka, A.; Szczesny, L.; Fryzowicz, A.; Maczynski, J.; Dworak, L.B. Isokinetic and Isometric Assessment of the Knee Joint Extensors and Flexors of Professional Volleyball Players. *Int. J. Environ. Res. Public Health* **2021**, *18*, 6780. [CrossRef]
32. Tanner, J.M. Normal growth and techniques of growth assessment. *Clin. Endocrinol. Metab.* **1986**, *15*, 411–451. [CrossRef] [PubMed]
33. Moore, S.A.; McKay, H.A.; Macdonald, H.; Nettlefold, L.; Baxter-Jones, A.D.; Cameron, N.; Brasher, P.M. Enhancing a somatic maturity prediction model. *Med. Sci. Sports Exerc.* **2015**, *47*, 1755–1764. [CrossRef] [PubMed]
34. Fuchs, P.X.; Fusco, A.; Cortis, C.; Wagner, H. Effects of differential jump training on balance performance in female volleyball players. *Appl. Sci.* **2020**, *10*, 5921. [CrossRef]
35. Baumbach, S.F.; Brumann, M.; Binder, J.; Mutschler, W.; Regauer, M.; Polzer, H. The influence of knee position on ankle dorsiflexion—A biometric study. *BMC Musculoskelet. Disord.* **2014**, *15*, 246. [CrossRef] [PubMed]
36. Papaiakovou, G.; Kollias, I.; Siatras, T.; Panoutsakopoulos, V. The ankle joint and its influence upon dynamic and kinematic characteristics in a standing vertical jump. *Exerc. Soc.* **2002**, *32*, 30–40.
37. Fuchs, P.X.; Menzel, H.K.; Guidotti, F.; Bell, J.; von Duvillard, S.P.; Wagner, H. Spike jump biomechanics in male versus female elite volleyball players. *J. Sports Sci.* **2019**, *37*, 2411–2419. [CrossRef]
38. Papaiakovou, G.; Katsikas, F.; Nikodelis, T.; Panoutsakopoulos, V.; Kollias, I. Influence of the ankle joint dorsiflexion on the execution of vertical jumps. In Proceedings of the XXIV International Symposium on Biomechanics in Sports, Salzburg, Austria, 14–18 July 2006; Schwameder, H., Strutzenberger, G., Fastenbauer, V., Lindinger, S., Muller, E., Eds.; University of Salzburg: Salzburg, Austria, 2006; Volume 1, pp. 448–451.
39. Winter, D.A. *Biomechanics and Motor Control of Human Movement*, 2nd ed.; John Willey & Sons, Inc.: Toronto, ON, Canada, 1990.
40. Ebben, W.P.; Petushek, E.J. Using the reactive strength index modified to evaluate plyometric performance. *J. Strength Cond. Res.* **2010**, *24*, 1983–1987. [CrossRef]
41. Pleša, J.; Kozinc, Ž.; Smajla, D.; Šarabon, N. The association between reactive strength index and reactive strength index modified with approach jump performance. *PLoS ONE* **2022**, *17*, e0264144. [CrossRef]
42. Hara, M.; Shibayama, A.; Takeshita, D.; Hay, D.C.; Fukashiro, S. A comparison of the mechanical effect of arm swing and countermovement on the lower extremities in vertical jumping. *Hum. Mov. Sci.* **2008**, *27*, 636–648. [CrossRef]
43. Feltner, M.E.; Fraschetti, D.J.; Crisp, R.J. Upper extremity augmentation of lower extremity kinetics during countermovement vertical jumps. *J. Sports Sci.* **1999**, *17*, 449–466. [CrossRef]
44. Harman, E.A.; Rosenstein, M.T.; Frykman, P.N.; Rosenstein, R.M. The effects of arms and countermovement on vertical jumping. *Med. Sci. Sports Exerc.* **1990**, *22*, 825–833. [CrossRef] [PubMed]
45. Walsh, M.S.; Böhm, H.; Butterfield, M.M.; Santhosam, J. Gender bias in the effects of arms and countermovement on jumping performance. *J. Strength Cond. Res.* **2007**, *21*, 362–366. [CrossRef] [PubMed]
46. Arakawa, H.; Nagano, A.; Hay, D.C.; Kanehisa, H. The effects of ankle restriction on the multijoint coordination of vertical jumping. *J. Appl. Biomech.* **2013**, *29*, 468–473. [CrossRef] [PubMed]
47. Rousanoglou, E.N.; Georgiadis, G.V.; Boudolos, K.D. Muscular strength and jumping performance relationships in young women athletes. *J. Strength Cond. Res.* **2008**, *22*, 1375–1378. [CrossRef]

48. Nikolaidis, P.T.; Ziv, G.; Arnon, M.; Lidor, R. Physical characteristics and physiological attributes of female volleyball players--the need for individual data. *J. Strength Cond. Res.* **2012**, *26*, 2547–2557. [CrossRef]
49. Lidor, R.; Ziv, G. Physical characteristics and physiological attributes of adolescent volleyball players-a review. *Pediatr. Exerc. Sci.* **2010**, *22*, 114–134. [CrossRef]
50. Barker, L.A.; Harry, J.R.; Mercer, J.A. Relationships between countermovement jump ground reaction forces and jump height, reactive strength index, and jump time. *J. Strength Cond. Res.* **2018**, *32*, 248–254. [CrossRef]
51. Kipp, K.; Kiely, M.T.; Geiser, C.F. Reactive Strength Index Modified is a valid measure of explosiveness in collegiate female volleyball players. *J. Strength Cond. Res.* **2016**, *30*, 1341–1347. [CrossRef]
52. McMahon, J.J.; Jones, P.A.; Suchomel, T.J.; Lake, J.; Comfort, P. Influence of the Reactive Strength Index Modified on force- and power-time curves. *Int. J. Sports Physiol. Perform.* **2018**, *13*, 220–227. [CrossRef]
53. Lees, A.; Vanrenterghem, J.; De Clercq, D. Understanding how an arm swing enhances performance in the vertical jump. *J. Biomech.* **2004**, *37*, 1929–1940. [CrossRef]
54. Konrad, A.; Reiner, M.M.; Bernsteiner, D.; Glashüttner, C.; Thaller, S.; Tilp, M. Joint flexibility and isometric strength parameters are not relevant determinants for countermovement jump performance. *Int. J. Environ. Res. Public Health* **2021**, *18*, 2510. [CrossRef] [PubMed]
55. Bénard, M.R.; Harlaar, J.; Becher, J.G.; Huijing, P.A.; Jaspers, R.T. Effects of growth on geometry of gastrocnemius muscle in children: A three-dimensional ultrasound analysis. *J. Anat.* **2011**, *219*, 388–402. [CrossRef] [PubMed]
56. Cho, K.H.; Jeon, Y.; Lee, H. Range of motion of the ankle according to pushing force, gender and knee position. *Ann. Rehabil. Med.* **2016**, *40*, 27127–27128. [CrossRef] [PubMed]
57. Panoutsakopoulos, V.; Kotzamanidou, M.C.; Giannakos, A.K.; Kollias, I.A. Relationship of vertical jump performance and ankle joint range of motion: Effect of knee joint angle and handedness in young adult handball players. *Sports* **2022**, *10*, 86. [CrossRef] [PubMed]
58. Panidi, I.; Bogdanis, G.C.; Terzis, G.; Donti, A.; Konrad, A.; Gaspari, V.; Donti, O. Muscle architectural and functional adaptations following 12-weeks of stretching in adolescent female athletes. *Front. Physiol.* **2021**, *12*, 701338. [CrossRef]
59. Lieber, R.L.; Ward, S.R. Skeletal muscle design to meet functional demands. *Philos. Trans. R. Soc. Lond B Biol. Sci.* **2011**, *366*, 1466–1476. [CrossRef]
60. Bishop, D. Warm up I: Potential mechanisms and the effects of passive warm up on exercise performance. *Sports Med.* **2003**, *33*, 439–454. [CrossRef]
61. Kubo, K.; Kanehisa, H.; Kawakami, Y.; Fukunaga, T. Influence of static stretching on viscoelastic properties of human tendon structures in vivo. *J. Appl. Physiol.* **2001**, *90*, 520–527. [CrossRef]
62. de Weijer, V.C.; Gorniak, G.C.; Shamus, E. The effect of static stretch and warm-up exercise on hamstring length over the course of 24 h. *J. Orthop. Sports Phys. Ther.* **2003**, *33*, 727–733. [CrossRef]
63. Markou, S.; Vagenas, G. Multivariate isokinetic asymmetry of the knee and shoulder in elite volleyball players. *Eur. J. Sport Sci.* **2006**, *6*, 71–80. [CrossRef]
64. Bishop, C.; Read, P.; Chavda, S.; Jarvis, P.; Brazier, J.; Bromley, T.; Turner, A. Magnitude or direction? Seasonal variation of interlimb asymmetry in elite academy soccer players. *J. Strength Cond. Res.* **2022**, *36*, 1031–1037. [CrossRef] [PubMed]

Journal of
Functional Morphology and Kinesiology

MDPI

Article

Inter-Device Reliability of a Three-Dimensional Markerless Motion Capture System Quantifying Elementary Movement Patterns in Humans

Nicolas M. Philipp *, Dimitrije Cabarkapa *, Damjana V. Cabarkapa, Drake A. Eserhaut and Andrew C. Fry

Jayhawk Athletic Performance Laboratory—Wu Tsai Human Performance Alliance, Department of Health, Sport and Exercise Science, Lawrence, KS 66045, USA; d927c184@ku.edu (D.V.C.); drake.eserhaut@ku.edu (D.A.E.); acfry@ku.edu (A.C.F.)
* Correspondence: nicophilipp@ku.edu (N.M.P.); dcabarkapa@ku.edu (D.C.)

Abstract: With advancements in technology able to quantify wide-ranging features of human movement, the aim of the present study was to investigate the inter-device technological reliability of a three-dimensional markerless motion capture system (3D-MCS), quantifying different movement tasks. A total of 20 healthy individuals performed a test battery consisting of 29 different movements, from which 214 different metrics were derived. Two 3D-MCS located in close proximity were utilized to quantify movement characteristics. Independent sample *t*-tests with selected reliability statistics (i.e., intraclass correlation coefficient (ICC), effect sizes, and mean absolute differences) were used to evaluate the agreement between the two systems. The study results suggested that 95.7% of all metrics analyzed revealed negligible or small between-device effect sizes. Further, 91.6% of all metrics analyzed showed moderate or better agreement when looking at the ICC values, while 32.2% of all metrics showed excellent agreement. For metrics measuring joint angles (198 metrics), the mean difference between systems was 2.9 degrees, while for metrics investigating distance measures (16 metrics; e.g., center of mass depth), the mean difference between systems was 0.62 cm. Caution is advised when trying to generalize the study findings beyond the specific technology and software used in this investigation. Given the technological reliability reported in this study, as well as the logistical and time-related limitations associated with marker-based motion capture systems, it may be suggested that 3D-MCS present practitioners with an opportunity to reliably and efficiently measure the movement characteristics of patients and athletes. This has implications for monitoring the health/performance of a broad range of populations.

Keywords: motion capture; biomechanics; markerless; movement; kinetic; kinematic; analysis

Citation: Philipp, N.M.; Cabarkapa, D.; Cabarkapa, D.V.; Eserhaut, D.A.; Fry, A.C. Inter-Device Reliability of a Three-Dimensional Markerless Motion Capture System Quantifying Elementary Movement Patterns in Humans. *J. Funct. Morphol. Kinesiol.* **2023**, *8*, 69. https://doi.org/10.3390/jfmk8020069

Academic Editor: Diego Minciacchi

Received: 14 April 2023
Revised: 12 May 2023
Accepted: 17 May 2023
Published: 22 May 2023

1. Introduction

In recent years, the world of sport and human movement has experienced a great increase in the use and availability of technological devices to quantify and analyze various features of human health and performance. Amongst those technologies are biomechanical devices such as force platforms, quantifying movement characteristics from a kinetic standpoint, as well as motion capture systems focusing primarily on the acquisition of kinematic data such as joint ranges of motion and segment alignment. As part of this recent increase in the availability of sports and human performance technologies, scientists, researchers, as well as bioengineers have continuously studied the ways in which such devices may be created to collect data in a more time-efficient and user-friendly manner. For instance, once strictly laboratory-based force platforms are now portable and allow practitioners to gather data on large numbers of athletes or patients in minimal time. Similarly, advances in motion capture technology that have typically required the placement of reflective markers on various anatomical landmarks on the human body have resulted in markerless motion capture systems that do not require such markers on the human body and therefore may have a greater practical application in sport and health settings [1,2]. Beyond that, previous research reports have reported that the location and movement of the skin on which markers are placed, relative to actual skeletal motion and location, may present challenges with regard to the acquisition of repeatable and valid data in humans [3,4]. Within marker-based motion capture, this phenomenon is termed a soft tissue artifact and can lead to inaccuracies in the estimation of rigid body poses or kinematics [5].

Largely due to the above-mentioned limitations of marker-based systems, markerless motion capture solutions are starting to be explored in clinical and rehabilitation settings, as well as athletic settings [6]. A recently published study suggested that functional movement screening scores, non-invasively derived from a three-dimensional markerless motion capture system (3D-MCS), may provide health and fitness practitioners with key insights into a range of physical fitness parameters [7]. Particular markerless motion capture systems even advertise their capability of simultaneously quantifying kinetic data such as ground reaction forces, in addition to kinematic data. In 2016, Fry et al. [8] reported that ground reaction forces may be accurately derived from a motion capture system, using inverse dynamics, without the use of force platforms. Similarly, Cabarkapa et al. [9,10] suggested the reliability of utilizing markerless motion capture technology for measuring the kinetic properties of the basketball dunk, as well as the repeatability of motion health screening scores derived from an identical system. Being able to glean kinetic as well as kinematic data from only one system without having to apply reflective markers, or the additional use of force platforms would allow practitioners to test individuals in a more time-efficient manner. Within a recent SWOT (i.e., strength, weakness, opportunity, and threat) analysis looking at portable and low-cost markerless motion capture systems, Armitano-Lago et al. [2] proposed that markerless motion capture systems show considerable promise with regard to enhancing our understanding of human movement characteristics, especially in providing unrestricted and simple movement assessments in natural sporting contexts. While still limited, a growing body of literature has proposed the validity of markerless motion capture systems when compared to marker-based systems [11–15]. For instance, Sandau et al. [11] suggested that a markerless motion capture system was able to reliably produce data within the sagittal and frontal plane of motion during walking (e.g., joint flexion, extension, abduction, and adduction). However, data in the transverse plane (e.g., internal rotation, external rotations, eversions, and inversions) were deemed to be less reliable compared to those of a marker-based motion capture system. Looking at sagittal plane kinematics in a vertical jump task, Drazan et al. [12] found a very strong agreement between a custom markerless model approach and a gold-standard marker-based system. Further, Schmitz et al. [15] reported small differences in accuracy and reliability between a marker-based system and a single-camera markerless motion capture system. On the other hand, Harsted et al. [16] found reliability scores that were moderately acceptable for most measures but unacceptable for knee valgus and varus when comparing a markerless motion capture system to a traditional marker-based system during jumping tasks in preschool children. Similarly, Hando et al. [17] used a markerless motion capture system to identify potential associations between movement screening composite scores of vulnerabilities and injury risk in military trainees. In this study, the markerless motion capture composite scores only displayed poor to moderate test–retest reliability and failed to demonstrate the ability to discriminate between individuals that did and did not suffer subsequent musculoskeletal injuries [17]. It should be noted that the injuries were not classified as contact or non-contact injuries, which makes interpretation of the data challenging [17]. Still in their infancy, other works in the existing literature within the field of health and sport have used markerless motion capture systems to gain insights into human health and movement characteristics [10,18–20]. While the previously highlighted literature suggests the potentially effective use of markerless motion capture technologies, particularly from a validity standpoint, certain degrees of uncertainty pertaining to the reliability of such devices still remain, especially across a wide range of movement tasks and variables. Additionally, the reliability of markerless motion capture includes both technological reliability (e.g., between-device agreement), as well as biological reliability, testing the ability of a human to adequately repeat the motions being tested.

With the previously highlighted evolution of innovative markerless motion capture systems in mind, the aim of the present study was to determine the inter-device reliability (i.e., technical reliability) between two identical markerless motion capture systems placed in close proximity to each other. We hypothesized that the two identical systems would reflect good technological reliability for a variety of different movement tasks and the respective joint and segmental angle variables and center of mass distance measures in healthy individuals. The authors see value in investigating the novel technologies' reliability (i.e., technical and biological) prior to comparison with established industry gold standards for validity purposes.

2. Materials and Methods

2.1. Participants

A total of 20 healthy men (n = 11, height = 181 ± 7.2 cm, body mass = 87.7 ± 11.1 kg, age = 26.8 ± 6.8 years) and women (n = 9, height = 167 ± 6.6 cm, body mass = 62.7 ± 6.9 kg, age = 24.2 ± 7.3 years) volunteered to participate in the study. Prior to any testing, the subjects completed a health history questionnaire, indicating they were free of musculoskeletal injury. All participants signed an informed consent form. All respective study procedures were approved by the University of Kansas's Institutional Review Board.

2.2. Procedures

As part of this study, all subjects performed a total of 29 different movements (Table 1). These procedures were preceded by a dynamic warm-up protocol that included cycling on a stationary bike for 5 min. Relevant kinematic data from these movements were quantified using two 3D-MCS (DARI Motion, Overland Park, KS, USA) composed of eight high-definition cameras recording at 60 fps. These cameras were attached to a metal frame surrounding the testing area. Corresponding cameras from each system were placed next to each other in close proximity. General data collection procedures for this study were adapted from Cabarkapa et al. (2022). The hull technology model records and subtracts the visual signal minus the background, which is used to generate a pixelated person in order to obtain biomechanical parameters of interest. Following manufacturer guidelines, each system was separately calibrated prior to testing. Specific movement tasks were explained and demonstrated by the principal investigator of the study. Following this demonstration, the member of the research team running the motion capture system provided the subject with the following command: "three, two, one, go". Following the "go" command, the subject completed the movement task which was being recorded by the two 3D-MCSs. After the completion of the respective movement task, the command "done" was provided to the subject, to indicate the end of the movement. Instructions for the completion of all 29 movements remained identical for all 20 subjects. A total of 214 metrics were derived from the movement battery, including 198 joint and segment angle variables, and 16 distance measures (e.g., center of mass movement).

2.3. Statistical Analysis

Prior to any analyses, all data were checked for normal distribution using Shapiro–Wilk's statistics. To determine between-device differences (System 1 vs. System 2), independent t-tests were used for all variables of interest. Data with a normal distribution were analyzed using Student's independent t-test, while the Mann–Whitney U statistical test was used for data that were not normally distributed. For the student's t-tests, mean and standard deviation values were reported, while the median was reported for the Mann–Whitney U tests. For parametric data, Cohen's d effect sizes were calculated and interpreted as negligible ($\leq$0.10), small (0.11–0.50), moderate (0.51–0.75), and large (>0.75) [21]. Effect sizes for non-parametric data were interpreted as described within the previous sentence, following a conversion from $\eta 2$ to Cohen's d [22,23]. Additionally, intraclass correlation coefficients (ICC) were used to examine the agreement between respective metrics of interest. ICCs were interpreted following suggestions by Koo and Li [24], where <0.50 was deemed poor reliability, 0.50–0.74 was deemed moderate reliability, 0.75–0.90 was deemed good reliability, and >0.90 was deemed excellent reliability. Lastly, mean absolute between-system differences were reported to indicate the actual difference between systems for distance measures (cm) and degrees (deg). All statistical inferences were made using an alpha level of <0.05. Data were analyzed using the R statistical computing environment and language (v. 4.0; R Core Team, 2020) via the Jamovi graphical user interface.

Table 1. List and description of all 29 tested movements.

Specific Movement Performed	Description of Movement
Shoulder Abduction	Start with arms at your sides with your palms facing forward. With arms straight, raise them out from your sides and over your head (abduct), keeping palms forward throughout the entire movement
Shoulder Horizontal Abduction	Start with your arms out in front of you at shoulder height with your palms facing each other. Bring your arms away from each other and behind your body as far as possible, keeping them at shoulder height throughout
Shoulder Internal/External Rotation	Start with elbows and shoulders bent at 90 degrees and palms facing down. Rotate arms up and back as far as possible (externally), and then forward and down (internally). Keeping elbows in the same spot during the movement
Shoulder Flexion/Extension	Begin with arms by your side. In one fluid motion, bring hands forward and up above the head, then down and back behind the body, and then return to original position.
Forward Fold	Begin with feet shoulder width apart. Tuck chin to chest and continue to round the back forward, bending at the hips in an attempt to touch the forehead to the knees.
Trunk Lateral Flexion Right	Begin with feet shoulder width apart and hands by the sides. Keep right hand on the outside of the right leg and bend upper body to the right as far down as possible, then return to starting position
Trunk Lateral Flexion Left	Begin with feet shoulder width apart and hands by the sides. Keep left hand on the outside of the right leg and bend upper body to the right as far down as possible, then return to starting position
Trunk Rotation	Start with elbows and shoulders bent at 90 degrees and palms facing down. In one fluid motion, rotate arms, torso, and head, first to the right, then to the left, and then return to starting position
Reverse Lunge with Rotation Right	Begin with arms out to the sides and elbows bent. Reach left leg back and drop into lunge without letting left knee touch the ground. At the bottom of the lunge, rotate the trunk to the right as far as possible, then return to starting position
Reverse Lunge with Rotation Left	Begin with arms out to the sides and elbows bent. Reach right leg back and drop into lunge without letting right knee touch the ground. At the bottom of the lunge, rotate the trunk to the left as far as possible, then return to starting position
Body Weight Squat	Begin with feet shoulder width apart and toes pointing forward. In one fluid motion, squat as low as possible, then return to the starting position
Overhead Squat	Begin with feet shoulder width apart, toes pointing forward and the dowel rod held above the head, with hands positioned wider than shoulders. In one fluid motion, squat as low as possible, and return to the original position
Forward Lunge Right	Begin by striding out with right leg getting as far and deep as possible. Then return to the starting position in one fluid motion. During movement keep arms out for balance
Forward Lunge Left	Begin by striding out with left leg getting as far and deep as possible. Then return to the starting position in one fluid motion. During movement keep arms out for balance
Lateral Lunge Right	Begin by stepping out as far to the right as possible. While allowing arms to travel out in front of the body, lunge as low as possible. Then return to the starting position
Lateral Lunge Left	Begin by stepping out as far to the left as possible. While allowing arms to travel out in front of the body, lunge as low as possible. Then return to the starting position

Table 1. *Cont.*

Specific Movement Performed	Description of Movement
Standing Hip Abduction Right	Begin with hands on hips, standing with feet together. Keep right leg straight and raise it out to the side as far as possible, then return to the starting position
Standing Hip Abduction Left	Begin with hands on hips, standing with feet together. Keep left leg straight and raise it out to the side as far as possible, then return to the starting position
Unilateral Squat Right	Transfer weight to the right leg, lifting the left foot off the ground and behind the body. In one fluid motion, squat as low as possible, keeping the left foot off the ground, and arms out for balance
Unilateral Squat Left	Transfer weight to the right leg, lifting the left foot off the ground and behind the body. In one fluid motion, squat as low as possible, keeping the left foot off the ground, and arms out for balance
Countermovement Vertical Jump	Begin by standing with feet shoulder width apart. Load and jump as high as possible. Do not step into the jump, but you may use an arm swing
Static Vertical Jump	Begin by standing with feet shoulder width apart. Lower into a squat position with arms repositioned to a natural jumping stance. Remain in this position for two seconds. On the signal "jump" immediately jump as high as possible from the squat position
Unilateral Vertical Jump Right	Begin by standing on right leg with left foot off the ground behind the body. Load and jump as high as possible, using an arm swing, and landing on your right foot again
Unilateral Vertical Jump Left	Begin by standing on left leg with left foot off the ground behind the body. Load and jump as high as possible, using an arm swing, and landing on your left foot again
Lateral Bound Right	Begin by taking two large steps to the left. Push off with the left leg and bound as far to the right side as possible. Land on the right leg and immediately push off in the opposite direction to reach the starting position
Lateral Bound Left	Begin by taking two large steps to the right. Push off with the right leg and bound as far to the left side as possible. Land on the left leg and immediately push off in the opposite direction to reach the starting position
5 Hop Right	Begin standing on the right leg with left foot off the ground behind the body. Jump on the right leg five times. Jump as high as possible, and as fast as possible, spending as little time on the ground between jumps as possible
5 Hop Left	Begin standing on the left leg with right foot off the ground behind the body. Jump on the left leg five times. Jump as high as possible, and as fast as possible, spending as little time on the ground between jumps as possible
Drop Jump	Begin standing on a 30-cm-high box. With either foot, step off the box landing on two feet. Immediately jump for maximal height, spending as little time as possible on the ground. An arm swing may be used

3. Results

Descriptive statistics for all variable comparisons may be found in the supplementary file (Table S1). Of the 214 variables reported, 94.9% of the metrics revealed negligible or small between-device effect sizes. Further, 91.6% of all metrics analyzed showed moderate or better agreement when looking at the ICC values, including 32.2% of all metrics showing excellent agreement. Only 2.3% of all metrics were significantly different when compared to the other system. Summary statistics for effect sizes and ICC values may be found in Tables 2 and 3, respectively. For metrics measuring joint angles, the mean absolute difference between systems was 2.9 degrees (198 metrics), while for metrics investigating distance measures (e.g., center of mass depth), the mean difference between systems was 0.62 cm (16 metrics).

Table 2. Summary statistics for effect sizes.

Effect Size	Number of Variables (% From Total)
Negligible (≤0.10)	72 variables (33.7%)
Small (0.11–0.50)	131 variables (61.2%)
Moderate (0.51–0.75)	11 variables (5.1%)
Large (>0.75)	0 variables (0.0%)

Table 3. Summary statistics for interclass correlations coefficients (ICC) values.

ICC	Metric Count (n = 214)	Total	Metric Count (% of Total)	Total (%)
≥0.90	69	-	32.2%	-
≥0.80–0.89	54	123	25.3%	57.5%
≥0.70–0.79	38	161	17.8%	75.2%
≥0.60–0.69	19	180	8.9%	84.1%
≥0.50–0.59	16	196	7.5%	91.6%
≥0.40–0.49	6	202	2.8%	94.4%
≥0.30–0.39	9	211	4.2%	98.6%
≥0.20–0.29	3	214	1.4%	100%
<0.20	0	-	-	-

4. Discussion

While previous research reports have investigated the biological reliability and validity of variables derived from 3D-MCSs [10,11,17], this study aimed to quantify technological reliability for the underlying variables from which all calculated measures are derived. More specifically, the aim of this study was to quantify the between-device agreement of two identical markerless motion capture systems located in close proximity. To the authors knowledge, this is the first study investigating the inter-device reliability of a markerless motion-capture system, capturing a plethora of elementary movements, from which a wide range of metrics are gleaned. Study findings revealed that a broad range of reliability scores (i.e., ICC or ES) were found across the selected metrics. Up to 75% of the metrics presented ICC scores of 0.70 or higher, reflecting moderate to excellent agreement. Similarly, when looking at effect sizes for between-system comparisons, 95.7% of all metrics suggested small or negligible effect sizes. Previous research reports looking at the reliability of markerless motion capture systems reported good reliability for movements performed within the sagittal or frontal plane, while movements performed within the transverse plane (e.g., rotations) revealed fewer stable measures [11]. Within our data, this suggestion is only partially reflected. When looking at internal and external shoulder rotation, all four metrics present ICC values of 0.90 or higher. For the trunk rotation exercise, ICC values for lumbar and thoracic rotation range from 0.69–0.81. Lastly, for the reverse lunge with rotation movement, ICC values for lumbar and thoracic rotation only range from 0.29 to 0.78. However, the metric presenting the 0.29 ICC value was accompanied by a small effect size. Our data suggest that reliability may not only be influenced by the plane in which movements are performed but also by the specific movements and body parts that are being investigated. Hip angles displayed mean absolute differences ranging from 0.3 degrees to 11.4 degrees, with the average being 6.3 degrees across 28 different hip flexion measures. While looking at the validity of a markerless motion capture system, Harsted et al. [16] reported moderate to poor agreement for a range of hip flexion measures extracted from squats and jumps, when comparing the markerless motion capture system to a marker-based system. In their study, between-system differences ranged from 5.8 degrees to 14.8 degrees [16].

A very commonly implemented and analyzed movement from a rehabilitation and athletic performance standpoint is the countermovement vertical jump [12]. While using a different analysis technique, Drazan et al. [12] found very strong agreement between a markerless motion capture technology and a gold-standard marker-based system when looking at a number of different angular measurements of the hip, knee, and ankle, for the vertical jump. In our study, moderate to excellent agreement between the two markerless motion capture systems were found for all countermovement vertical jump metrics, except for ankle flexion and knee valgus during the eccentric phase of the jump, and upon landing. Similarly, for the drop jump, ICC values for ankle flexion and dynamic valgus during landing only ranged from 0.29 to 0.49, with effect sizes of 0.10 to 0.62. All kinematic measures within the countermovement vertical and drop jump presented considerably more between-system agreement. It should be noted that many of the valgus measures were made during activities not typically performed for clinical assessments and must be interpreted accordingly. When looking at the upper body movements, 9 out of 12 metrics presented ICC values of good to excellent agreement. One should be cognizant that the shoulder flexion metrics presenting with lower ICC values showed non-significant differences, and small to negligible effect sizes, suggesting small to negligible between-system differences. Distance measures such as the center of mass depth during different movement tasks such as vertical jumps have been of interest to practitioners working in sport performance settings. For instance, Merrigan et al. [25] suggested that the center of mass depth during a countermovement vertical jump was an important contributor to overall jumping capability. Within our data, distance measures such as center of mass depths were in good agreement between the two systems, indicated by only two metrics presenting ICC values under 0.90, and only one metric presenting an ICC value under 0.85. In dynamic movements such as countermovement vertical jumps, the center of mass moves through a significant range of motion, making it important to acquire a reliable measure of center of mass movements. Within marker-based systems, markers attached to the skin using double adhesive tape can influence normal movement patterns and can move relative to the underlying bone, commonly known as a skin movement artifact or soft tissue artifact [3–5]. Beyond that, certain movements, clothes worn, or ranges of motion can even lead to further issues with regard to these reflective markers. A recent study even had to standardize the squat depth of the participants in order to avoid the occlusion of the reflective markers on the anterior superior illicit spine at lower squat depths [26]. This could certainly restrict an individual's normal range of motion, posing limitations to the acquisition of practically relevant data. Further, this may influence not only the efficiency of assessment procedures but also potentially the reliability of derived metrics.

Readers should be cognizant of the potential limitations when interpreting the results of this study. From a procedural perspective, given the broad range of movement tasks, participants only performed one repetition for each movement, which is in line with the suggestions for test administration given by the manufacturer of the motion capture system. Future studies may aim to narrow in on the specific movements identified within this study, investigating aspects of reliability across additional repetitions. Pending appropriate reliability studies, technologies such as the one used in this study should further be compared to marker-based tracking systems to gain added insights into the validity of markerless motion capture systems. Additionally, the breadth and depth of the movements and respective metrics did not allow the authors to provide a detailed discussion of all variables. Furthermore, the landscape of available markerless motion capture solutions is rapidly expanding, making it difficult to generalize findings from this study across other systems and software. Lastly, the group of participants consisted entirely of healthy individuals. Assuming that healthy individuals present less variability when performing the movement battery, future investigations may aim to replicate study procedures within injured or rehabilitating groups. This may lend insights into the between-device reliability of markerless motion capture systems across a broader range of movement characteristics. Especially in return-to-play scenarios for athletes a more frequent evaluation of movement characteristics could greatly aid clinical practitioners in evaluating patients' progress as well as in their ability to individually tailor a specific return-to-play protocol [18]. In many cases, this assessment frequency is hindered by the cumbersome and time-consuming nature of marker-based motion capture systems. Therefore, 3D-MCSs could greatly enhance the work of a broad range of practitioners as well as the health and performance of different groups of individuals.

5. Conclusions

The study findings suggested promising results with regard to the between-device reliability of a 3D-MCS quantifying a broad range of movements. Overall, nearly all of the variables assessed demonstrated acceptable to strong inter-device reliability, indicating low technological variability. Readers are encouraged to employ caution when trying to generalize the results of this study past the specific system and software used. Given the technological reliability reported in this study as well as the logistical and time-related limitations associated with marker-based motion capture systems, it is concluded that this specific markerless motion capture technology presents practitioners with an opportunity to measure the movement characteristics of patients and athletes at a greater frequency, due in part to its high technological reliability. This could have implications for the health and performance of a broad range of populations.

Supplementary Materials: The following supporting information can be downloaded at: https://www.mdpi.com/article/10.3390/jfmk8020069/s1, Table S1: Comparison data for inter-device reliability (technical reliability) for two identical markerless motion capture systems. Results are presented as $\bar{x} \pm$ SD for normally distributed data, or medians for non-normally distributed data. Significance (p-value) for individual t-tests mean absolute differences ($\bar{x}$-diff), effect sizes (ES), and intra-class correlation coefficients (ICC) are also listed for each variable.

Author Contributions: Conceptualization, N.M.P., D.C. and A.C.F.; methodology, N.M.P., D.C. and A.C.F.; formal analysis, N.M.P. and A.C.F.; investigation, N.M.P., D.C., D.V.C., D.A.E. and A.C.F.; data curation, N.M.P. and A.C.F.; writing—original draft preparation, N.M.P. and D.C.; writing—review and editing, N.M.P., D.C., D.V.C., D.A.E. and A.C.F. All authors have read and agreed to the published version of the manuscript.

Funding: This research received no external funding.

Institutional Review Board Statement: The study was conducted according to the guidelines of the Declaration of Helsinki and approved by the Institutional Review Board, #00148349 (8 March 2022).

Informed Consent Statement: Written informed consent was obtained from all participants involved in the study.

Data Availability Statement: The data presented in this study are available upon request from the corresponding author.

Acknowledgments: The authors would like to thank Isabella Martinoli for her help in collecting the data used in this study. Further, the authors would like to thank the Wu Tsai Human Performance Alliance Foundation for their help in conducting this research project.

Conflicts of Interest: The authors declare no conflict of interest.

References

1. Mündermann, L.; Corazza, S.; Andriacchi, T.P. The evolution of methods for the capture of human movement leading to markerless motion capture for biomechanical applications. *J. Neuroeng. Rehabil.* **2006**, *3*, 6. [CrossRef] [PubMed]
2. Armitano-Lago, C.; Willoughby, D.; Kiefer, A.W. A SWOT analysis of portable and low-cost markerless motion capture systems to assess lower-limb musculoskeletal kinematics in sport. *Front. Sports Act. Living* **2021**, *3*, 405. [CrossRef] [PubMed]
3. Reinschmidt, C.; van den Bogert, A.J.; Nigg, B.M.; Lundberg, A.; Murphy, N. Effect of skin movement on the analysis of skeletal knee joint motion during running. *J. Biomech.* **1997**, *30*, 729–732. [CrossRef] [PubMed]
4. Morton, N.A.; Maletsky, L.P.; Pal, S.; Laz, P.J. Effect of variability in anatomical landmark location on knee kinematic description. *J. Orthop. Res.* **2007**, *25*, 1221–1230. [CrossRef] [PubMed]
5. Ancillao, A.; Aertbeliën, E.; De Schutter, J. Effect of the soft tissue artifact on marker measurements and on the calculation of the helical axis of the knee during a gait cycle: A study on the CAMS-Knee data set. *Hum. Mov. Sci.* **2021**, *80*, 102866. [CrossRef]
6. Wade, L.; Needham, L.; McGuigan, P.; Bilzon, J. Applications and limitations of current markerless motion capture methods for clinical gait biomechanics. *PeerJ* **2002**, *10*, e12995. [CrossRef]

7. Cabarkapa, D.; Whetstone, J.M.; Patterson, A.M.; Mosier, E.M.; Cabarkapa, D.V.; Fry, A.C. Relationship between health-related physical fitness parameters and functional movement screening scores acquired from a three-dimensional markerless motion capture system. *Int. J. Environ. Res. Public Health* **2022**, *19*, 4551. [CrossRef]
8. Fry, A.C.; Herda, T.J.; Sterczala, A.J.; Cooper, M.A.; Andre, M.J. Validation of a motion capture system for deriving accurate ground reaction forces without a force plate. *Big Data Anal.* **2016**, *1*, 1–8. [CrossRef]
9. Cabarkapa, D.; Fry, A.C.; Mosier, E.M. Validity of 3-D markerless motion capture system for assessing basketball dunk kinetics—A case study. *Sport J.* **2020**. Available online: https://thesportjournal.org/article/tag/basketball/ (accessed on 7 May 2023).
10. Cabarkapa, D.; Cabarkapa, D.V.; Philipp, N.M.; Downey, G.G.; Fry, A.C. Repeatability of motion health screening scores acquired from a three-dimensional markerless motion capture system. *J. Funct. Morph Kinesiol.* **2022**, *7*, 65. [CrossRef]
11. Sandau, M.; Koblauch, H.; Moeslund, T.B.; Aanæs, H.; Alkjær, T.; Simonsen, E.B. Markerless motion capture can provide reliable 3D gait kinematics in the sagittal and frontal plane. *Med. Eng. Phys.* **2014**, *36*, 1168–1175. [CrossRef]
12. Drazan, J.F.; Phillips, W.T.; Seethapathi, N.; Hullfish, T.J.; Baxter, J.R. Moving outside the lab: Markerless motion capture accurately quantifies sagittal plane kinematics during the vertical jump. *J. Biomech.* **2021**, *125*, 110547. [CrossRef]
13. Tanaka, R.; Ishikawa, Y.; Yamasaki, T.; Diez, A. Accuracy of classifying the movement strategy in the functional reach test using a markerless motion capture system. *J. Med. Eng. Technol.* **2019**, *43*, 133–138. [CrossRef]
14. Wochatz, M.; Tilgner, N.; Mueller, S.; Rabe, S.; Eichler, S.; John, M.; Völler, H.; Mayer, F. Reliability and validity of the Kinect V2 for the assessment of lower extremity rehabilitation exercises. *Gait. Posture* **2019**, *70*, 330–335. [CrossRef]
15. Schmitz, A.; Ye, M.; Shapiro, R.; Yang, R.; Noehren, B. Accuracy and repeatability of joint angles measured using a single camera markerless motion capture system. *J. Biomech.* **2014**, *47*, 587–591. [CrossRef]
16. Harsted, S.; Holsgaard-Larsen, A.; Hestbæk, L.; Boyle, E.; Lauridsen, H.H. Concurrent validity of lower extremity kinematics and jump characteristics captured in pre-school children by a markerless 3D motion capture system. *Chirop. Man. Ther.* **2019**, *27*, 39. [CrossRef]
17. Hando, B.R.; Scott, W.C.; Bryant, J.F.; Tchandja, J.N.; Scott, R.M.; Angadi, S.S. Association between markerless motion capture screenings and musculoskeletal injury risk for military trainees: A large cohort and reliability study. *Orthop. J. Sport. Med.* **2021**, *9*, 23259671211041656. [CrossRef]
18. Daggett, M.C.; Witte, K.A.; Cabarkapa, D.; Cabarkapa, D.V.; Fry, A.C. Evidence-based data models for return-to-play criteria after anterior cruciate ligament reconstruction. *Healthcare* **2022**, *10*, 929. [CrossRef]
19. Mosier, E.M.; Fry, A.C.; Nicoll, J.X.; Cabarkapa, D.; Sontag, S.A. Test-retest reliability of performance scores using a markerless motion capture system. *Int. J. Exerc. Sci. Conf. Proc.* **2019**, *11*, 53. [CrossRef]
20. Bird, M.B.; Koltun, K.J.; Mi, Q.; Lovalekar, M.; Martin, B.J.; Doyle, T.L.A.; Nindl, B.C. Predictive utility of commercial grade technologies for assessing musculoskeletal injury risk in US Marine Corps Officer candidates. *Front. Physiol.* **2023**, *14*, 25. [CrossRef]
21. Cohen, J. *Statistical Power Analysis for the Behavioral Sciences*, 2nd ed.; Erlbaum: Hillsdale, NJ, USA, 1988.
22. Fritz, C.O.; Morris, P.E.; Richler, J.J. Effect size estimates: Current use, calculations, and interpretation. *J. Exp. Psychol.* **2022**, *141*, 2–18. [CrossRef]
23. Lenhard, W.; Lenhard, A. Computation of effect sizes. *Psychometrica* **2016**. Available online: https://www.psychometrica.de/effect_size.html (accessed on 5 May 2023).
24. Koo, T.K.; Li, M.Y. A guideline of selecting and reporting intraclass correlation coefficients for reliability research. *J. Chiropr. Med.* **2016**, *15*, 155–163. [CrossRef] [PubMed]
25. Merrigan, J.J.; Rentz, L.E.; Hornsby, W.G.; Wagle, J.P.; Stone, J.D.; Smith, H.T.; Galster, S.M.; Joseph, M.; Hagen, J.A. Comparisons of countermovement jump force-time characteristics among National Collegiate Athletic Association Division I American football athletes: Use of principal component analysis. *J. Strength Cond. Res.* **2022**, *36*, 411–419. [CrossRef]
26. Al-Amri, M.; Nicholas, K.; Button, K.; Sparkes, V.; Sheeran, L.; Davies, J.L. Inertial measurement units for clinical movement analysis: Reliability and concurrent validity. *Sensors* **2018**, *18*, 719. [CrossRef]

Journal of
Functional Morphology and Kinesiology

MDPI

Article

Effects of Age and Popularity of Sport on Differences among Wrestlers' Parental Support: An Exploratory Study

Ivica Biletic [1], Hrvoje Karnincic [2,*] and Mario Baic [3]

[1] Police Academy—The First Croatian Police Officer, University of Applied Sciences in Criminal Investigation and Public Security, 10000 Zagreb, Croatia; biletic@net.hr
[2] Faculty of Kinesiology, University of Split, 21000 Split, Croatia
[3] Faculty of Kinesiology, University of Zagreb, 10000 Zagreb, Croatia; mario.baic@kif.unizg.hr
* Correspondence: hrvojek@kifst.hr; Tel.: +385-91-556-0730

Abstract: No research was previously performed on wrestling related to parental support. It is not known whether there are differences in support between younger and older children. The popularity of a sport can be reflected in parental support, and parents may be more inclined towards popular sports. The aim of this research was to examine differences in parental support among wrestlers of different age categories and between those coming from communities in which wrestling is a popular sport versus communities in which it is less popular. The sample of participants consisted of 172 wrestlers. The Parental Support Scale for Children in Sports was applied. Parental willingness to set an example was lower. As far as age is concerned, the period of entry into specialisation is sensitive. At this age, children perceive less parental support ($p = 0.04$) and lower parental belief in the benefits of sports ($p = 0.01$). The popularity of the sport is related to parental support. In environments in which wrestling is popular, parents know the sport better and can participate; therefore, children perceive more parental support. The findings of this study may help coaches to better understand athlete–parent relationships.

Keywords: age categories; parental involvement; combat sport

Citation: Biletic, I.; Karnincic, H.; Baic, M. Effects of Age and Popularity of Sport on Differences among Wrestlers' Parental Support: An Exploratory Study. *J. Funct. Morphol. Kinesiol.* **2023**, *8*, 65. https://doi.org/10.3390/jfmk8020065

Academic Editor: Diego Minciacchi

Received: 29 March 2023
Revised: 27 April 2023
Accepted: 4 May 2023
Published: 15 May 2023

1. Introduction

Parental support is an important factor in children's adherence to participation in sports [1], and it can also be a reason for practicing sports [2]. One of the most important socialising forces for children's entry into sports is the family, especially parents [3,4]. Parental beliefs are often related to their children's understanding of sport [5–7]. Woolgar and Power (1993) were the first to define three basic forms of parental involvement in children's sport: emotional support, the provision of information, and concrete help [8]. Bosnar (2003) identified two other important aspects: setting an example for children and positive reinforcement, which are related to expectations for children's sport performances. Expectations should be realistic (neither too low nor too high) to avoid having a negative effect on children's motivation [9]. Parental support for children in sports was researched through many concepts in numerous studies. Most generally speaking, all these studies and their findings can be reduced to a twofold concept: parental support and parental pressure. In the present research, the topic is parental support. The analysis of parental support can reveal some causal mechanisms behind children either continuing or quitting practicing sports. A better understanding of these mechanisms can increase the number of children engaged in sports. In an attempt to better investigate this topic, a series of questionnaires was created: the Parental Support Scale [10,11], Parental Involvement in Sport Questionnaire (PISQ) [12], Parent-Initiated Motivational Climate Questionnaire [13], and others. All these questionnaires have different foci; therefore, the findings or results of the studies that used them cannot be compared. What are comparable are general conclusions. The Croatian version of the questionnaire was designed by Bosnar and associates in 2003, and it was validated in samples of primary and secondary school children, as well as in samples of children involved in combat sports [9,14–16].

Giving up further involvement in sports is negatively correlated with the age of the athlete [17]. It is well known that growing up brings more responsibilities and less free time for training. The question arises as to whether parental support is one of the factors that causes older children to withdraw from sports. The popularity of a sport can be both positively and negatively reflected in parental support. Children's success in sports can change their parents' social status [18]. One of the reasons why children play sports is because they are popular, and children who play sports are more popular [18]. It is logical that this phenomenon is more pronounced in popular sports, in which the profits are higher. This can lead to negative consequences, such as an excessive focus on results. Fortunately, such parents are in the minority [19]. Problems with illegal substances in young athletes occur in popular sports for this reason; however, this is not yet a mass phenomenon [20]. Research showed that the most important thing for a man's popularity is sport, while for women, the most important thing is appearance [21]. In this paper, the focus is on the popularity of a sport in a particular setting. The popularity of a sport may vary in environments that are culturally and environmentally close to each other. For example, football, which is most popular in Europe, is not the most popular sport in the United States (where it is behind basketball, baseball, and American football) [22]. Parents who set overly ambitious goals and participate too much in their children's sports careers have a larger problem if the sport is popular, and especially if the earnings are high [23]. In previous research, it was demonstrated that parents of higher-standard athletes provide them with more autonomy in their upbringing, whereas the upbringing of lower-standard athletes is characterised by the excessive involvement of their parents [24].

According to the list of the Croatian Wrestling Federation, there are 30 active wrestling clubs in Croatia. In its capital city, Zagreb, and close surrounding areas, there are 14 wrestling clubs (nine in the city and five in the surrounding areas), together with numerous sections in schools. The rest of the 16 clubs are active in 14 Croatian cities. Only two cities recently established second wrestling clubs. Research showed that, in Australia, more children participate in sports where sports facilities are most accessible [25]. This is also the case in Croatia because the number of sports facilities for wrestling in the capital is much greater. However, in the same study, it was found that there were fewer sports facilities and less interest in sports in the capital [25]. In this study, the situation is reversed: half of all clubs and wrestlers in the country are located in the capital. We can assume that this difference is due to the different sizes of the countries. In the Croatian Premier First Wrestling League, six clubs are competing, of which four are from Zagreb. We can feasibly say that one-half of the Croatian wrestling activities are related to the capital city and its surroundings. Moreover, the longest tradition of wrestling is also related to the capital city—as many as six of the first wrestling clubs in Croatia were founded in Zagreb [26], and almost all of the important national and international competitions take place there. Therefore, we can say that, in the capital of Croatia, a larger number of people understand and follow the sport. The numerical indicators of the popularity of a sport are the number of spectators at sport events, its TV popularity ratings, and the number of sport clubs and athletes [27]. By all the criteria, wrestling in the capital is much more popular than in other Croatian cities.

The basic hypothesis is that there are differences in parental support between children of different ages, which is also logical from the perspective of the parent–child relationship in childhood or adolescence. Adolescence is a transitional phase to adulthood in which children seek as much autonomy as possible [28]. Another hypothesis is that there are differences in parental support in environments in which wrestling is popular and those in which it is not. The expectation is that some parents prefer to support their children when the sport is popular.

The aim of the research was to examine differences in parental support among wrestlers of different age categories, and between those coming from communities in which wrestling is a popular sport versus those in which it is less popular. The findings should provide new insights into the support of parents for young athletes.

2. Materials and Methods

2.1. Sample of Participants

The sample of participants consisted of 172 male wrestlers from 20 Croatian wrestling clubs. Several subsamples were created by the participants' ages and whether they were residents of the capital or another Croatian city. We can say that the sample was representative because it consisted of about 60% of young wrestlers of this age from more than 90% of the wrestling clubs in the country. Seven female wrestlers filled in the questionnaire; however, they were not part of the sample because there were too few of them and they were not foreseen in the study project. The exclusion of girls from the study was due to the short tradition of female wrestling in Croatia [29]. The sample included young female wrestlers who had been wrestling for at least one year. Because wrestling training starts at the age of 10 years [30], the youngest wrestlers in the sample were 11 years old. In accordance with the International Wrestling Rules, the following age categories were considered: precompetitors (n = 53; age: 11.58 ± 0.50 years), boys U15 (n = 75; age: 14.00 ± 0.84 years), and cadets (n = 44; age: 16.29 ± 0.46 years). In the age category of boys U15, children of 14 and 15 years of age have the right to compete; however, this right may also be granted to children of 13 years subject to the permission of their parents. Wrestlers from 9 wrestling clubs from the capital city were represented in the sample (n = 93; age: 14.15 ± 1.83 years), whereas 11 clubs were from other Croatian cities (n = 79; age: 13.42 ± 1.85 years).

2.2. Sample of Variables

Participant responses to 25 items of the Parental Support Scale for Children in Sports [9] were the variables of our research. The questionnaire was originally designed in the Croatian language and was validated in samples of individual and team sports [9,14,31], as well as in samples of younger-aged athletes [31,32]. The questionnaire consisted of four subscales: 1. Parental beliefs in benefits of doing sports (9 items: 1, 3, 9, 11, 13, 16, 17, 19, and 25 (example: My parents think it is important for me to do sports to be healthier)); 2. Ensuring material conditions for doing sports (6 items: 6, 7, 14, 15, 22, and 23 (example: My parents pay for my extracurricular sports activities)); 3. Learning from role models (3 items: 4, 12, and 20 (example: My family and I often do a sporting activity together, regardless of our age differences)); 4. Positive reinforcement (7 items: 2, 5, 8, 10, 18, 21, and 24 (example: When talking about me to other people, my parents like to point out that I play sports)). Respondents were presented with 25 statements and were required to answer to what extent they agreed with them. Responses to the 25 items on the questionnaire were provided on a five-point Likert scale ranging from "strongly agree" to "strongly disagree". The research used unpublished data from the doctoral dissertation Social Environment and Youth Participation in Wrestling [33]. The research was conducted in compliance with the Declaration of Helsinki and was approved by the Ethical Committee of the Faculty of Kinesiology, University of Zagreb (approval number: 57/2019).

2.3. Data Processing Methods

Data were processed by the program package Statistica for Windows, version 14.0 (TIBCO, Software Inc., Palo Alto, CA, USA). To test the reliability of the scale (questionnaire), Cronbach's alpha and interitem correlation (IIC) were computed. All the variables were processed by descriptive statistics (median, mode, and mode frequency), whereas for the subscales, the mean and standard deviation were computed. The differences among the subscales were determined using the Friedman ANOVA and Wilcoxon matched-pairs test. For the determination of the differences among age groups, the Kruskal–Wallis test was used, and for the differences between the capital and other cities, the Mann–Whitney U test and Wilcoxon matched-pairs test were utilised. The level of statistical significance was set at $p < 0.05$.

2.4. Research Protocol

After the study was approved by the administrators of the wrestling clubs, parental consent was obtained. Test dates were then arranged with the clubs. The clubs were asked to provide a room where the children could complete the questionnaire at their leisure. The young wrestlers filled out the questionnaire anonymously and without the presence of a coach in a quiet place in the club rooms before or after training in groups of from five to ten. Because wrestlers of different age groups train at different times, the data were separately collected by age group. The children filled out the questionnaire with a pen on paper, as the youngest group required additional explanations that would not have been possible with the online method. It took approximately 15–30 min to complete the questionnaire. The method for completing the questionnaire was explained to all participants in the same way, and the research supervisor was available to clarify any ambiguities related to it. In each session, the principal investigator informed the children that they did not have to answer a question if it made them uncomfortable. Data collection took place at the beginning of the competition season. Data were collected within one month in all cities in which there were wrestling schools. Incorrectly completed questionnaires (many missing answers, the same answer for all questions) were removed from the sample (out of 196, 172 were correct).

3. Results

From Table 1, it is obvious that the reliability parameters (Cronbach's alpha and IIC) indicate a high reliability of the measuring instrument in both the entire sample of participants and in each subsample.

Table 1. Questionnaire reliability parameters (Cronbach's alpha, interitem correlation (IIC)) for all participants and all groups separately.

	N	Cronbach's Alpha	IIC
All groups	172	0.87	0.25
Precompetitors	53	0.81	0.19
Boys	79	0.83	0.21
Cadets	44	0.90	0.30
Capital city	93	0.87	0.25
Other cities	76	0.88	0.25

Note. *N*: number of respondents; IIC: interitem correlation.

Significant age differences (Table 2) were obtained in Subscale 1 (Parental beliefs in benefits of doing sports) and Subscale 4 (Positive reinforcement from parents). Moreover, a significant difference was obtained between the milieus (wrestling is popular vs. wrestling is less popular) in Subscale 4 (Positive reinforcement from parents).

Table 2. Descriptive statistical parameters (mean and standard deviation) for all subscales and the differences among them (Friedman ANOVA, Wilcoxon matched-pairs test).

	Precompetitors	Boys	Cadets	Friedman ANOVA	Capital	Other Cities	Wilcoxon Matched Pairs Test
Subscales	Mean ± SD	Mean ± SD	Mean ± SD	(ANOVA Chi sqr.)	Mean ± SD	Mean ± SD	
Subscale 1	4.34 ± 1.01	4.16 ± 1.10	4.36 ± 0.95	Chi sqr. = 6.22 $p = 0.04$	4.36 ± 0.99	4.24 ± 1.03	Z = 1.60 $p = 0.11$
Subscale 2	4.57 ± 0.79	4.53 ± 0.84	4.53 ± 0.82	Chi sqr. = 2.33 $p = 0.31$	4.55 ± 0.78	4.53 ± 0.86	Z = −0.31 $p = 0.75$
Subscale 3	3.82 ± 1.13	3.56 ± 1.33	3.88 ± 1.25	Chi sqr. = 4.66 $p = 0.09$	3.71 ± 1.19	3.74 ± 1.32	Z = −0.00 $p = 1.00$
Subscale 4	4.59 ± 0.72	4.36 ± 0.94	4.63 ± 0.63	Chi sqr. = 10.57 $p < 0.01$	4.52 ± 0.80	4.48 ± 0.82	Z = 1.18 $p = 0.02$

Note. Subscale 1: Parental beliefs in benefits of doing sports; Subscale 2: Ensuring material conditions for doing sports; Subscale 3: Learning from role models; Subscale 4: Positive reinforcement from parents.

Table 3 reveals statistically significant age differences in the responses to Items 8 (My parents are proud of me doing sports (p = 0.02)), 9 (My parents want me to do sports so that I would develop agility and strength (p = 0.02)), 16 (When they talk to other people about me, my parents are happy to point out that I do sports (p = 0.05)), and 18 (My parents are happy when my sport teammates meet up at our place ($p \leq 0.01$)). Graphs of significantly different variables can be found in the Supplementary Files (Figures S1–S4).

Table 3. Descriptive statistical parameters (median, mode, mode frequency) for all three age groups and the differences among them (Kruskal–Wallis ANOVA).

	Precompetitors (n = 53)			U 15 (n = 75)			Cadets (n = 44)			Kruskal–Wallis ANOVA	
Item	Median	Mode	F Mode	Median	Mode	F Mode	Median	Mode	F Mode	H	p
1	5	5	47	5	5	58	5	5	34	4.35	0.11
2	5	5	45	5	5	56	5	5	34	1.42	0.49
3	4	4	17	3	3	32	4	5	17	3.78	0.15
4	4	3	18	3	3	21	4	5	17	1.09	0.58
5	5	5	38	5	5	47	5	5	26	5.46	0.07
6	5	5	49	5	5	59	5	5	34	1.09	0.58
7	5	5	43	5	5	63	5	5	37	7.42	0.02 *
8	5	5	48	5	5	60	5	5	39	7.42	0.02 *
9	5	5	46	5	5	52	5	5	33	4.24	0.12
10	5	5	43	5	5	56	5	5	38	0.42	0.81
11	4	5	23	4	5	37	4	5	21	1.77	0.41
12	4	5	23	4	5	33	5	5	26	4.00	0.10
13	5	5	30	4	5	32	5	5	25	0.02	0.99
14	5	5	38	5	5	53	5	5	29	5.95	0.75
15	5	5	40	5	5	57	5	5	30	5.95	0.05 *
16	5	5	39	5	5	48	5	5	35	5.81	0.06
17	5	5	38	5	5	43	5	5	31	12.27	<0.01 *
18	4	5	25	4	5	29	5	5	29	0.44	0.80
19	5	5	35	5	5	51	5	5	29	1.31	0.52
20	4	5	18	4	5	28	4	5	15	4.92	0.09
21	5	5	33	4	5	36	5	5	23	1.09	0.58
22	5	5	35	5	5	46	5	5	28	0.48	0.79
23	5	5	33	5	5	46	5	5	22	1.36	0.51
24	5	5	38	5	5	48	5	5	29	2.36	0.31
25	5	5	36	5	5	42	5	5	26	4.35	0.11

* Statistically significant difference between groups.

In Table 4, statistically significant differences are obvious between the wrestlers from the capital and those from other Croatian cities in the responses to Item 9 (My parents think athletes are a good company for me to keep (p = 0.04)) and Item 15 (My sports equipment was financed by my parents (p = 0.03)). Graphs of significantly different variables can be found in the Supplementary Files (Figures S5 and S6).

Table 4. Descriptive statistical parameters (median, mode, mode frequency) and differences (Mann–Whitney U test) between wrestlers from the capital city and those from other cities.

	Capital City (n = 93)			Other Cities (n = 79)			Mann–Whitney U Test		
Item	Median	Mode	F Mode	Median	Mode	F Mode	U	Z	p-Value
1	5	5	79	5	5	60	3370.50	0.93	0.35
2	5	5	75	5	5	60	3485.00	0.44	0.66
3	3	3	38	4	5	27	3341.50	−1.02	0.31
4	4	Multiple	25	4	5	28	3493.00	−0.55	0.58
5	5	5	60	5	5	51	3625.00	0.00	1.00
6	5	5	76	5	5	66	3616.50	−0.17	0.86
7	5	5	80	5	5	63	3417.00	0.79	0.43
8	5	5	80	5	5	67	3629.00	0.14	0.89
9	5	5	79	5	5	52	3010.50	2.04	0.04 *
10	5	5	77	5	5	60	3374.00	0.92	0.36
11	5	5	47	4	5	34	3143.00	1.63	0.10
12	4	5	45	4	5	37	3459.50	0.66	0.51
13	5	5	54	4	5	33	3110.00	1.73	0.08
14	5	5	65	5	5	55	3669.00	0.01	0.99
15	5	5	77	5	5	50	2948.50	2.23	0.03 *
16	5	5	68	5	5	54	3505.00	0.52	0.61
17	5	5	61	5	5	51	3596.00	−0.24	0.81
18	5	5	51	4	5	32	3208.50	1.43	0.15
19	5	5	68	5	5	47	3118.50	1.70	0.09
20	4	5	29	4	5	32	3296.50	−1.16	0.25
21	5	5	53	4	5	39	3557.00	0.36	0.72
22	5	5	54	5	5	55	3090.00	−1.66	0.10
23	5	5	55	5	5	46	3612.50	−0.19	0.85
24	5	5	60	5	5	55	3455.00	−0.67	0.50
25	5	5	58	5	5	46	3564.00	0.33	0.74

* Statistically significant difference between groups.

4. Discussion

The study was conducted with the aim of determining the differences in parental support between wrestlers of different age groups, and between those from communities in which wrestling is a popular sport and those in which it is less popular. This research demonstrated that age and popularity were related to parental support, which was determined as remarkably high. Comparable results (i.e., high parental support) were recorded in other studies as well [9,16]. Analysing the differences between the age groups, we found that there were significant differences in Subscale 1 (Parental beliefs in benefits of doing sports) and Subscale 4 (Positive reinforcement from parents). It is interesting that the group of boys U15 scored lower than both the younger (precompetitors) and older (cadets) wrestlers, which can be explained by their developmental stage. At this stage, boys commence their phase of specialisation; wrestling-specific contents are more focused on training, and first combat experiences are gathered. This is the stage at which children want more parental praise and understanding [34]. We can assume that, for the same reason, the boys U15 responses were lower on Subscale 1 as well (Parental beliefs in benefits of doing sports). First combat experiences can lead to fears and dilemmas for parents. The involvement of parents in their children's sports depends not only on their gender (mother or father), sports experience, or lack of sports experience, but also on the child's current stage of sports development [35]. Comments are very important for children at this stage in order for them to properly perceive their competence in sports, and their self-confidence depends on this. Children should not only receive comments from primary sources (coaches, teammates), but also from secondary sources (friends, parents) [36]. Comments are very important for children at this stage to properly assess their competence in sports, and their self-confidence depends on this. To support their children in sports, parents need to know something about the developmental stages in sports: "Before kids can play like a pro, they need to enjoy playing the game like a kid," said Steve Locker, an international soccer player and coach [36]. Parental support should not turn into parental pressure, and the line is very thin. Although this is not an article about parental pressure, we must note this very thin line. Research showed that parents who are prone to pressure are willing to push their children towards early athletic specialisation, which is not good for either their mental or physical development [36–38]. Positive reinforcement from parents to children is higher in milieus in which wrestling is popular. The popularity of the sport had no effect on the other subscales. Parental comments on children's sports depend, to a large extent, on whether the parents feel competent enough to comment on them [39]. In milieus in which wrestling is popular, more people practice it and/or follow wrestling competitions. In these communities, there are many more competent people who have no problem commenting on their children's sport. Therefore, it was logical to expect that a lower level of positive support for children in sports would be present in the milieus in which wrestling is less popular. In Subscale 3 (Learning from role models/parents), no significant difference was found between the groups; however, we should emphasise here that the scores on this scale were low. Such a finding causes concern because the lack of positive role models was demonstrated to be one of the main reasons that children give up sports [40]. Similar findings were obtained by Crnjac in 2017. In the sample of 472 adolescent athletes from combat sports, the scores were between 4.06 and 4.48 for Subscales 1, 2, and 4, whereas the score for Subscale 3 was 3.57 [15]. Unfortunately, results on problematic Subscale 3 (Learning from role models/parents) cannot be compared with the findings of other studies because, apart from the Parent Support Scale Questionnaire [11], in which the Sports Habits of Parents subscale is included, no one has specifically analysed this problem, to the best of the authors' knowledge. Sage (1980) conducted similar research and demonstrated that the variable Time spent participating with their offspring was lower and weakly related to socioeconomic status [41]. We can assume that more time spent with children means more opportunities to offer role model examples. All research conducted with the Parental Support Scale for Children in Sports questionnaire [9] points to the same problem: parents should demonstrate more by their own example. In our research as well, in all the groups, the same problem was ascertained as far as parents were concerned—a very high willingness to encourage children to participate in sports, a high willingness to finance children's engagement in sports, and the attitude that sports are beneficial for their children, but not a very high willingness to show children parental attitudes towards sports by example. In this study, there was no age difference in Subscale 3 (Learning from role models/parents). This is interesting because it was expected that the oldest group would have a lower score on this scale. At their age, the adjustment to adulthood takes place [28]. This is the time when children run away from anything that they believe might hinder their transition to adulthood. The overinvolvement of parents in sports may also be a reason why young people give up sports [28].

If individual items of the questionnaire are analysed separately, then age differences occurred in 4 out of 25 items, whereas differences between the groups by wrestling popularity were recorded in two items. In Items 8 (My parents are proud of me doing sports), 9 (My parents want me to do sports so that I would develop agility and strength), and 16 (When they talk to other people about me, my parents are happy to point out that I do sports), the U15 wrestlers differed from the wrestlers of both the younger and older age groups. Children usually commence practicing wrestling at the age of 10 years [30,42]. The youngest group in our research had an average age of 11.5 years. The children should be at the stage of multilateral sports development and not yet at the specialisation stage; thus, they have not yet started seriously competing. Wrestlers of the second age group (U15) are participating in their first serious matches, whereas the cadets already gathered some competitive experience. From this aspect, the U15 wrestlers and their parents have the most dilemmas; children have their first combats but do not have any previous experience as to whether is it safe or how dangerous it is. Parents may have different perceptions of injuries and safety risks and encourage children to play sports [43]. Anxiety is a normal phenomenon in combat sports [44], but it is reduced with experience [44,45]. In Item 18 (My parents are happy when my sport teammates meet up at our place), the youngest group of wrestlers differed from the other two age groups. Our assumption is that that the reason for this finding is not the sport itself but the children's age—at this age, they are too young to wonder across cities alone, someone needs to bring them to meet their friends, and they need special attention, which is not the case with the older age groups.

Similar to age differences, there were no differences between wrestlers coming from different milieus in terms of wrestling popularity in the majority of the questionnaire items except for two: My parents think athletes are a good company for me to keep and My sports equipment was financed by my parents. It is interesting that in the cities where wrestling is popular and widespread, parents have a dilemma as to whether colleagues from sports are good company for their children. We can assume that the dilemma arises from the fear of parents related to the violent behaviour of the young. It is obvious that parents have fears related to the safety of their children, and we can assume that this phenomenon is more prevalent in environments in which there are more wrestlers and wrestling clubs [43]. However, previous research on bullying demonstrated that the athletes from combat sports are not more often culprits of bullying or other aggressive behaviours than athletes from other sports [46]. In fact, most violent behaviour and aggression was committed by team sport athletes, whereas combat sport athletes have less often been victims compared with athletes from other sports in Croatia [46,47], where wrestling is most popular in its capital—Zagreb. Moreover, in the capital, citizens' personal incomes are, on average, higher than in other parts of Croatia [48]. Therefore, it is feasible to assume that the higher personal incomes are the reason that the citizens of Zagreb are more willing to finance their children's sports than their peers from other parts of Croatia.

5. Conclusions

This research was conducted with the aim of examining parental support for young wrestlers and to ascertain the possible relationships between this support, wrestlers' ages, and the popularity of wrestling in the milieus from which they came. Parental support for young wrestlers was very high; on a scale from 1 to 5, the scores were in the range of from 4.3 to 4.5 for Subscales 1, 2, and 4, whereas Subscale 3 (Learning from role models/parents) was rated 3.75 on average. Parents believe that the sport is good for their children, they are ready to finance it, and they support their children in it; however, their willingness to show by example is lower. The analysis by age showed that when children enter the specialisation stage of their sports development, they seek more support from their parents, while parents have concerns about their children's first combats. The popularity of the sport is related to parental support. In communities in which wrestling is more popular, parents better understand and know the sport, and so, they can participate in commenting on it and directly support their children. Moreover, our research showed that the parents in areas with higher personal incomes more willingly spend more money on their children's sport, which is not necessarily related to the popularity of the sport.

Parents should become better acquainted with their children's sports so that they can provide better support and comment on them together with their children, and they should lead by their own example. Coaches are the ones who could suggest these things to parents so that the parent–child bond in sports can grow stronger. However, caution is always recommended because there is a very thin line between parental support and parental pressure.

Study Limitation

Several studies already dealt with this problem; however, the authors were mainly focused on the creation of new tools. Therefore, the results of these studies cannot be numerically compared but can only be commented on and compared in general because of the different methodologies used. Future studies should also include a sample of female wrestlers, whose numbers significantly increased and are sufficient for research. Our assumption is that parental support is not the same for males and females.

Supplementary Materials: The following supporting information can be downloaded at: https://www.mdpi.com/article/10.3390/jfmk8020065/s1.

Author Contributions: Conceptualization, I.B. and M.B.; methodology, H.K.; software, I.B. and H.K.; validation, I.B. and M.B.; formal analysis, I.B. and H.K.; investigation, I.B.; data curation, I.B.; writing—original draft preparation, I.B. and H.K.; writing—review and editing, H.K.; visualization, M.B. and I.B.; supervision, M.B.; project administration, I.B. All authors have read and agreed to the published version of the manuscript.

Funding: This research received no external funding.

Institutional Review Board Statement: The study was conducted in accordance with the Declaration of Helsinki, and was approved by the Ethical Committee of the Faculty of Kinesiology, University of Zagreb (the number of the approval: 57/2019).

Informed Consent Statement: Not applicable.

Data Availability Statement: Data and other supplementary files can be found at: https://drive.google.com/drive/folders/1rtnk4GU_8aGZBVJyiqrOKG3zbMvUSD17?usp=share_link (accessed on 1 April 2023).

Conflicts of Interest: The authors declare no conflict of interest.

References

1. Lindner, J. Withdrawal from competitive youth sport: A retrospective ten-year study. *J. Sport Behav.* **2002**, *25*, 7–8.
2. Eime, R.M.; Harvey, J.T.; Craike, M.J.; Symons, C.M.; Payne, W.R. Family support and ease of access link socio-economic status and sports club membership in adolescent girls: A mediation study. *Int. J. Behav. Nutr. Phys. Act.* **2013**, *10*, 50. [CrossRef] [PubMed]
3. Stroot, S.A. Socialisation and participation in sport. In *Sociology of Sport and Physical Education*; Routledge: London, UK, 2012; pp. 129–147.
4. Howard, D.R.; Madrigal, R. Who makes the decision: The parent or the child? The perceived influence of parents and children on the purchase of recreation services. *J. Leis. Res.* **1990**, *22*, 244–258. [CrossRef]
5. Leff, S.S.; Hoyle, R.H. Young athletes' perceptions of parental support and pressure. *J. Youth Adolesc.* **1995**, *24*, 187–203. [CrossRef]
6. Brustad, R.J. Attraction to physical activity in urban schoolchildren: Parental socialization and gender influences. *Res. Q. Exerc. Sport* **1996**, *67*, 316–323. [CrossRef]
7. Babkes, M.L.; Weiss, M.R. Parental influence on children's cognitive and affective responses to competitive soccer participation. *Pediatr. Exerc. Sci.* **1999**, *11*, 44–62. [CrossRef]
8. Woolger, C.; Power, T.G. Parent and sport socialization: Views from the achievement literature. *J. Sport Behav.* **1993**, *16*, 171–190.
9. Bosnar, K. Roditelji kao partneri školskog sporta: Empirijska mjera roditeljske potpore. In Proceedings of the V. Konferencija o Sportu Alpe-Jadran, Zagreb, Croatia, 5–7 June 2003; Puhak, S., Kristić, K., Eds.; Ministarstvo Prosvjete i Športa RH: Zagreb, Croatia, 2003; pp. 11–18.
10. Marcen, C.; Gimeno, F.; Gómez, C.; Sáenz, A.; Gutiérreza, H. Socioeconomic status, parental support, motivation and self-confidence in youth competitive sport. *Procedia Soc. Behav. Sci.* **2013**, *82*, 750–754. [CrossRef]
11. Zheng, N.; Yang, Y. Development and validation of parental support scale of children's sports training in China: Socio-ecological approach. *Front. Psychol.* **2022**, *13*, 922282. [CrossRef]
12. Özgör, N.; Özdemir, E.; İbanoğlu, E.; Koruç, Z.; Aşçi, F. Parental Involvement in Sport Questionnaire"(PISQ): Reliability and Validity study on Turkish Adolescent Athletes. In Proceedings of the IIV International Exercise and Sport Psychology Congress, Istanbul, Türkiye, 17 April 2018.
13. White, S.A.; Duda, J.L.; Hart, S. An exploratory examination of the parent-initiated motivational climate questionnaire. *Percept. Mot. Ski.* **1992**, *75*, 875–880. [CrossRef]

14. Bosnar, K.; Turkalj, I. Roditeljska potpora dječacima u individualnim i ekipnim sportovima. *Zagreb Kineziol. Fak. Sveučilišta U Zagreb.* **2011**, *11*, 145–153.
15. Crnjac, D. *Social Environment and Participation of Older Adolescents in Some Martial Arts*; Sveučilište u Zagrebu, Kineziološki Fakultet: Zagreb, Croatia, 2017.
16. Lorger, M.B.K. Mjerna svojstva skale roditeljske potpore u sportu učenika osnovne škole. In Proceedings of the 15 Summer School of Kinesiologists of the Republic of Croatia, Quality of Work in the Fields of Education, Sports and Sports Recreation, Rovinj, Croatia, 20–24 May 2006; pp. 162–167.
17. Zarotis, G.; Athanailidis, I.; Arvanitidou, V.; Mourtzios, C. Age-specific reasons for dropping out of the Fitness-Sport. *J. Phys. Educ. Sport* **2017**, *17*, 916.
18. White, S.A.; Duda, J.L.; Keller, M.R. The relationship between goal orientation and perceived purposes of sport among youth sport participants. *J. Sport Behav.* **1998**, *21*, 474.
19. Gould, D.; Lauer, L.; Rolo, C.; Jannes, C.; Pennisi, N. The role of parents in tennis success: Focus group interviews with junior coaches. *Sport Psychol.* **2008**, *22*, 18–37. [CrossRef]
20. Dandoy, C.; Gereige, R.S. Performance-enhancing drugs. *Pediatr. Rev.* **2012**, *33*, 265. [CrossRef]
21. Chase, M.A.; Dummer, G.M. The role of sports as a social status determinant for children. *Res. Q. Exerc. Sport* **1992**, *63*, 418–424. [CrossRef]
22. De Block, A.; Dewitte, S. Darwinism and the cultural evolution of sports. *Perspect. Biol. Med.* **2009**, *52*, 1–16. [CrossRef]
23. Siekańska, M. Athletes' perception of parental support and its influence on sports accomplishments—A retrospective study. *Hum. Mov.* **2012**, *13*, 380–387. [CrossRef]
24. Coutinho, P.; Ribeiro, J.; da Silva, S.M.; Fonseca, A.M.; Mesquita, I. The influence of parents, coaches, and peers in the long-term development of highly skilled and less skilled volleyball players. *Front. Psychol.* **2021**, *12*, 667542. [CrossRef]
25. Mei-Dan, O. Children and extreme sports: A parent's perspective. *Res. Sport. Med.* **2018**, *26*, 1–4. [CrossRef]
26. Marić, J.B.; Cvetković, Č.M. *Primjena Hrvanja u Ostalim Sportovima*; Faculty of Kinesiology, University of Zagreb: Zagreb, Croatia, 2007.
27. Varmus, M.; Kubina, M.; Boško, P.; Mičiak, M. Application of the perceived popularity of sports to support the sustainable management of sports organizations. *Sustainability* **2022**, *14*, 1927. [CrossRef]
28. Coakley, J.; White, A. Making decisions: Gender and sport participation among British adolescents. *Sociol. Sport J.* **1992**, *9*, 20–35. [CrossRef]
29. Biletić, I. *Socijalna Okolina i Sudjelovanje Mladih u Hrvanju*; University of Zagreb, Faculty of Kinesiology: Zagreb, Croatia, 2022.
30. Baić, M.; Karninčić, H.; Šprem, D. Beginning age, wrestling experience and wrestling peak performance—Trends in period 2002–2012. *Kinesiology* **2014**, *46*, 95–101.
31. Grebalo, Z.; Lorger, M. Roditeljska potpora djeci u momčadskim i ostalim sportovima. In Proceedings of the Ljetna Škola Kineziologa Republike Hrvatske "Kvaliteta Rada u Područjima Edukacije, Sporta i Sportske Rekreacije", Rovinj, Croatia, 20–24 June 2006; pp. 115–120.
32. Matković, S.P.; Bosnar, K.F. Roditeljska potpora mladim sportašima. In Proceedings of the V. Konferencija o sportu Alpe–Jadran, Rovinj, Croatia, 5–7 July 2003; pp. 121–127.
33. Biletić, I. *Social Environment and Youth Participation in Wrestling*; University of Zagreb, Faculty of Kinesiology: Zagreb, Croatia, 2022.
34. Zisi, V.; Giannitsopoulou, E.; Kosmidou, E. Examination of parental involvement in Greek female athletes. *J. Hum. Sport Exerc.* **2010**, *5*, 176–187.
35. Kovács, K.; Földi, R.F.; Gyömbér, N. The Effects of Parents' and Their Children's Characteristics on Parental Involvement in Sport. *J. Fam. Issues* **2022**, 0192513X221106739. [CrossRef]
36. Horn, T.S. Social psychological and developmental perspectives on early sport specialization. *Kinesiol. Rev.* **2015**, *4*, 248–266. [CrossRef]
37. Jayanthi, N.; Schley, S.; Cumming, S.P.; Myer, G.D.; Saffel, H.; Hartwig, T.; Gabbett, T.J. Developmental training model for the sport specialized youth athlete: A dynamic strategy for individualizing load-response during maturation. *Sport. Health* **2022**, *14*, 142–153. [CrossRef]
38. Normand, J.M.; Wolfe, A.; Peak, K. A review of early sport specialization in relation to the development of a young athlete. *Int. J. Kinesiol. Sport. Sci.* **2017**, *5*, 37–42. [CrossRef]
39. Holt, N.L.; Tamminen, K.A.; Black, D.E.; Sehn, Z.L.; Wall, M.P. Parental involvement in competitive youth sport settings. *Psychol. Sport Exerc.* **2008**, *9*, 663–685. [CrossRef]
40. Ciocan, D.M.; Milon, A. Study Regarding the Motives of Children Participation in Sport Activities. *LUMEN Proc.* **2017**, *1*, 159–168.
41. Sage, G.H. Parental influence and socialization into sport for male and female intercollegiate athletes. *J. Sport Soc. Issues* **1980**, *4*, 1–13. [CrossRef]
42. Baić, M.; Karninčić, H.; Mirzaei, B. Important Age Parameters in Wrestling—The Differences Between the European and World Trends. In Proceedings of the XI International Conference of Young Scientists "Youth and the Olympic Movement", Kyiv, Ukraine, 10–12 April 2018.
43. Boufous, S.; Finch, C.; Bauman, A. Parental safety concerns—A barrier to sport and physical activity in children? *Aust. N. Z. J. Public Health* **2004**, *28*, 482–486. [CrossRef]

44. Vaccaro, C.A.; Schrock, D.P.; McCabe, J.M. Managing emotional manhood: Fighting and fostering fear in mixed martial arts. *Soc. Psychol. Q.* **2011**, *74*, 414–437. [CrossRef]
45. Karninčić, H.; Baić, M.; Slačanac, K. Mood aspects of rapid weight loss in adolescent wrestlers. *Kinesiology* **2016**, *48*, 229–236. [CrossRef]
46. Marracho, P.; Pereira, A.M.A.; da Graça Nery, M.V.; Rosado, A.F.B.; de Castro, E.M.R.T. Is young athletes' bullying behaviour different in team, combat or individual sports? *Motricidade* **2021**, *17*, 70–78.
47. Vveinhardt, J.; Fominiene, V.B. Prevalence of bullying and harassment in youth sport: The case of different types of sport and participant role. *J. Hum. Sport Exerc.* **2022**, *17*, 272–292. [CrossRef]
48. Karaman Aksentijević, N.; Denona Bogović, N. Economic inequality and the influence of salaries on income inequality in the Republic of Croatia. *Zb. Rad. Ekon. Fak. U Rijeci Časopis Za Ekon. Teor. I Praksu* **2003**, *21*, 37–51.

Journal of
Functional Morphology and Kinesiology

MDPI

Article

New Specific Kinesthetic Differentiation Tests for Female Volleyball Players: Reliability, Discriminative Ability, and Usefulness

Karla Đolo , Zoran Grgantov and Goran Kuvačić *

Faculty of Kinesiology, University of Split, 21000 Split, Croatia; karla.djolo@gmail.com (K.Đ.); zoran.grgantov@kifst.eu (Z.G.)
* Correspondence: goran.kuvacic@kifst.eu

Abstract: This study aimed to determine the test-retest reliability and discriminative ability of five sport-specific kinesthetic differentiation ability tests in female volleyball players. The sample of participants consisted of 98 female volleyball players aged 15.20 $\pm$ 1 years from six clubs in Bosnia and Herzegovina. Kinesthetic differentiation ability was determined by the overhead passing test, forearm passing test, float service with a net test, float service without a net test, and float service 6 m from the net test. To estimate test-retest reliability, a sub-sample of 13 players performed all tests on two testing occasions. Furthermore, the discriminative ability of the tests was determined by analyzing the performance between players of different playing positions and situational performances. Parameters of the intraclass correlation coefficient (ICC) were excellent (0.87–0.78) in all tests except for the float service with the net test, whose reliability was good (0.66). For the absolute reliability estimates, the SEM was higher than SWC (0.2) for all variables except the float service 6 m from the net test, and the SEM was lower than SWC (0.6, 1.2). One-way ANOVA detected no statistically significant inter-positional differences in all five tests ($p > 0.05$). A significant difference was found between less and more successful players ($p < 0.01$) for all applied tests. The results of this study show that a specific battery test is a reliable and valid measure and can be used to monitor kinesthetic differentiation ability in young female volleyball players.

Keywords: coordination abilities; inter-positional difference; volleyball techniques

Citation: Đolo, K.; Grgantov, Z.; Kuvačić, G. New Specific Kinesthetic Differentiation Tests for Female Volleyball Players: Reliability, Discriminative Ability, and Usefulness. *J. Funct. Morphol. Kinesiol.* **2023**, *8*, 63. https://doi.org/10.3390/jfmk8020063

Academic Editor: Diego Minciacchi

Received: 27 March 2023
Revised: 8 May 2023
Accepted: 9 May 2023
Published: 12 May 2023

1. Introduction

Modern volleyball is a dynamic sport that involves multidirectional accelerations from landing to jumping, sudden starts and stops, changes of direction, attack, and defensive actions that require high levels of technical and tactical skills [1–3]. To achieve all these skills at a young age, training must be based on coordination abilities to excel in all game segments [4]. Otherwise, it is unlikely for athletes to reach their full potential later in life. Most studies emphasize the importance of sensitive phases as critical periods of enhanced children's coordination development [5,6]. In contrast, others doubt the existence of these phases and conclude that motor learning capacity can also improve in adolescence and later [7–9]. Nowadays, all researchers must keep in mind that a high level of skills cannot be possible if a child has not reached neural maturity [10].

Previous research highlights the significance of improving technical and tactical skills by applying all five basic coordination abilities: kinesthetic differentiation, spatial orientation, rhythm, balance, and reaction abilities [11,12]. Moreover, supplementing general coordination abilities with specific coordination will facilitate the learning process [13–15]. A competitive sport such as volleyball is challenging to master due to many technical aspects that require accuracy, such as serving, passing, and setting [16]. All of them represent the so-called ball feeling, partner feeling, and opponent feeling [17], which we can link with the most important coordination ability—kinesthetic differentiation. Researchers define it as a sense of upper and lower extremities and body movement where kinesthetic memory helps players to remember their motor movements, such as the skill needed for receiving service from different angles [7,18,19]. To successfully master this ability, players need to go through a large number of repetitions, monitored constantly until they no longer need to think about the movement; they just "do it" [20].

Until recently, many researchers have focused on constructing and developing tests to assess volleyball's general kinesthetic differentiation of lower and upper limbs [21–24]. General coordination tests, such as kinesthetic differentiation, are suitable for children before playing volleyball or for beginner volleyball players. They cannot perform specific tests because they have not yet perfected volleyball techniques. Therefore, it is common to apply general tests for jumping, throwing, and similar skills for players up to 14 years old. However, children aged 14 to 15 have already mastered volleyball techniques enough to be able to use specific coordination tests.

Recent studies have noted the need for sport-specific tests, which have contributed to modern sports and are more appropriate than standard tests because they are similar to real sports situations [25]. Nevertheless, there are insufficient reports on specific kinesthetic differentiation ability, especially in volleyball [1,26,27]. In short, these researchers focused on measuring skills to identify talented volleyball players by applying specific tests.

Specific volleyball coordination tests can be used as a means of talent identification to monitor the progress of volleyball players over a certain period. We can assume that more talented players will adopt volleyball techniques more rapidly, progress faster over time, and reach a higher level of skill [28]. Another possibility is to use them as a tool to assess the current level of volleyball technique. Therefore, the study aimed to examine the test-retest reliability and discriminative ability of new sport-specific kinesthetic differentiation ability tests in female volleyball players.

2. Materials and Methods

2.1. Study Design

This study had two stages: (1) test-retest reliability and (2) discriminative ability assessment. To assess the test-retest reliability, a subgroup of 13 participants underwent the same tests on two separate occasions, with a seven-day interval between them. The discriminative ability of the tests was determined by analyzing the group differences among players of different playing positions and situational performances.

During the investigation days, the players were instructed to avoid any factors that could hinder their effort during testing. This included refraining from consuming caffeine-containing beverages and low-fiber diets 24 h before and during the investigation in order to minimize any interference with the testing. Additionally, the athletes were instructed to refrain from engaging in physical activity 48 h before and during the investigation. To prevent circadian variations, all tests were conducted in the morning, starting at 8 am [29].

The testing was conducted in June 2021 with an ambient temperature ranging from 21 to 24 °C. At the beginning of the testing session, a standardized warm-up consisting of 10 min of jogging and mobility exercises was conducted. Players performed three maximum trials, followed by six trials of each skill. All tests were demonstrated and conducted by professionals, members of the Faculty of Kinesiology, University of Split, who introduced them to all testing procedures. Participants were given a briefing on the testing procedures prior to beginning any of the tests.

2.2. Participants

The study was conducted on 98 young female volleyball players, aged 15.20 ± 1.00 years, from six clubs in Bosnia and Herzegovina. The players specified their playing positions, which included setter (n = 17), passer-hitter (n = 35), opposite hitter (n = 16), middle blocker (n = 19), and libero (n = 11). Coaches from six volleyball clubs agreed to participate and received a clear explanation of the study. All players had been involved in at least two years of training. Informed consent was obtained from parents before players were permitted to participate. The inclusion criteria for players were being at a stage of specialized basic training with no musculoskeletal or psychophysical disorder. A professional volleyball coach supervised and regulated young athletes' training programs, which mainly comprised technical and tactical volleyball training and some physical conditioning sessions. On average, athletes trained for approximately 2 h per session, four days per week.

The situational performance of the players was assessed by combining 2 criteria [1]: (1) Placement of teams in the competition (Super League and 1st League); (2) Quality of individual players within their team where the coaches categorize their team players into 3 groups: group 1 including leading team players; group 2 including the rest of starting players and players entering the game, thus contributing to team result; and group 3 including players who very rarely or never enter the game. As presented in Table 1, a combination of these assessment criteria, each player is scored 1–5. More successful players were those that scored 5 and 4, and less successful were the ones that scored 3, 2, and 1. Therefore, the sample was divided into two subsamples: less successful (n = 59) and more successful (n = 39) players.

Table 1. The situational performance of the players according to the two criteria.

Team Placement	Within Team Players' Quality Evaluated by the Coach		
	Group 1	Group 2	Group 3
Super League	5	4	3
1st League	3	2	1

During the experiment, eight players who did not have an assigned playing position were excluded. Additionally, three players were excluded because they could not perform the service tests.

2.3. *Experimental Procedures*

The kinesthetic differentiation ability was assessed using five different volleyball skill tests: forearm passing, overhead passing, and three different standing float service tests, all on an indoor floor surface.

2.3.1. Forearm Passing Test (FPT)

In the first task, the athlete stands in front of a line with a ball in their hands (Figure 1). Their task is to throw the ball in front of themselves and forearm pass the ball three times as far as possible. A measuring scale is placed on the floor surface, and the ball bounces along that measuring scale. The measurer counts 50% of the athlete's maximum pass and marks that target with a cone. The kinesthetic differentiation during forearm passing is evaluated in the second part of the test, where athletes need to pass the ball as close to the cone as possible. The measurer records six attempts. Therefore, the test results represent the sum of deviations in centimeters in those six attempts (absolute values regardless of negative sign).

2.3.2. Overhead Pass Test (OPT)

The athlete begins by holding the ball and throwing it overhead to themselves (Figure 2). They then play the ball with an overhead pass as far as possible three times. The measuring scale is also placed on the floor surface, and the ball bounces along that measuring scale. After the athlete completes the three maximum passes, the measurer counts 50% of the maximum distance and marks that place with a cone. The kinesthetic differentiation during overhead passing is evaluated by asking athletes to overhead pass the ball as close to the cone as possible. The measurer records six attempts, and the test results represent the sum of deviations in centimeters in those six attempts (absolute values regardless of negative sign).

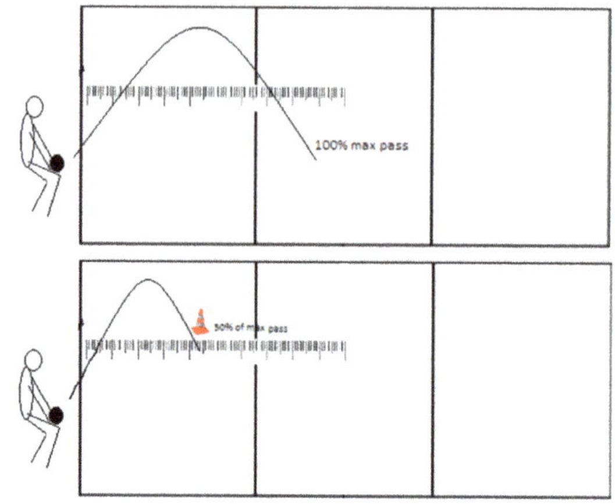

Figure 1. Visual representation of forearm passing test.

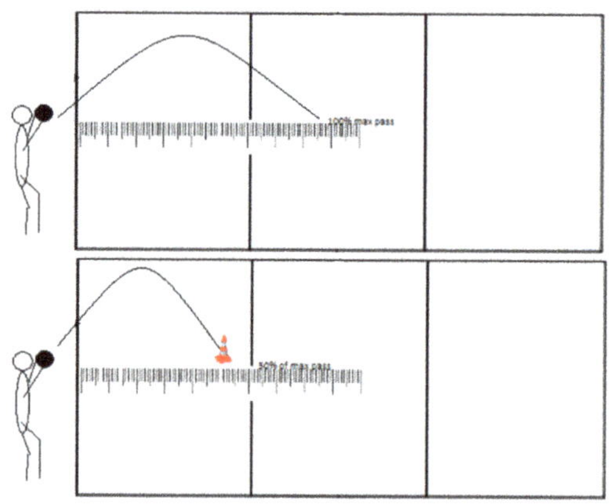

Figure 2. Visual representation of overhead passing test.

2.3.3. Float Service Tests (FST)

The standing float service accuracy was evaluated in three different tests: (1) float service with a net (FSTnet), (2) float service without a net (FSTx), and (3) float service 6 m from the net (FST6m). In all three tests, athletes first had to serve three maximum attempts on the court from a service position, followed by six attempts aiming at a target. Therefore, the test results represent the sum of deviations in centimeters in those six attempts (absolute values regardless of the negative sign).

In the FSTnet, athletes stand behind the baseline while serving (9 m from the net; Figure 3). After three maximum services over the net, the measurer counts 75% of the athlete's maximum serve and marks the target with a cone. Then, athletes have to serve the ball six times to that target. A measure of 75% was taken, unlike other tests, because the target would be farther from the net with this percentage. With 50%, the target for most participants would be near the net, and the test could not be performed since the net would interfere with hitting the target.

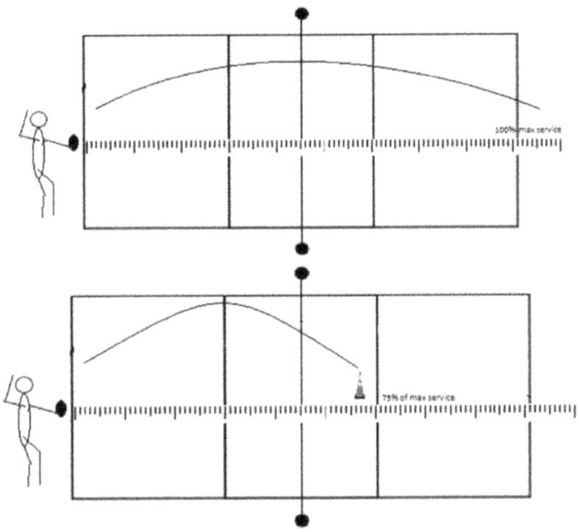

Figure 3. Visual representation of float service with a net test.

In the FSTx, the athlete serves in front of the baseline, imagining the net is on the court (Figure 4). The measurer counts 50% of the maximum serve and places the cone on that spot. Then, athletes have to serve the ball six times to that target.

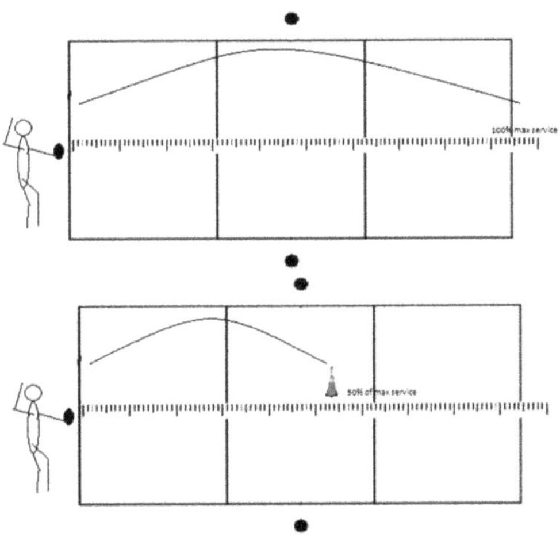

Figure 4. Visual representation of float service without a net test.

The third version of the test (FST6m) differs by positioning the athlete in front of the net after three maximum serves (Figure 5). The athlete serves three maximum serves from behind the baseline. Afterward, the athlete moves six meters from the net, from where they will serve the next six attempts. The measurer counts 50% of the maximum service and measures that distance, 6 m from the net. The athletes need to serve on the marked place over the net.

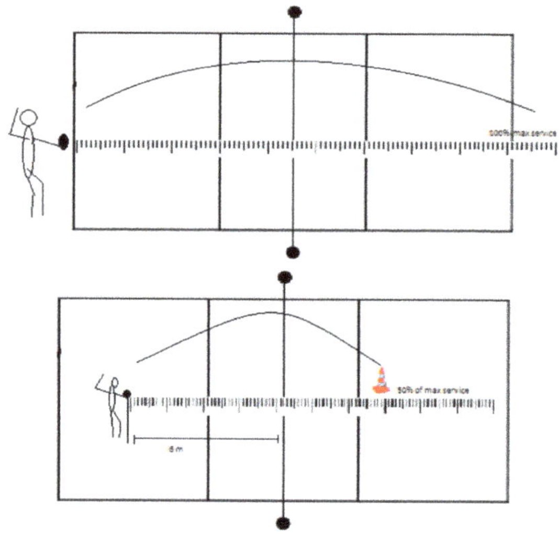

Figure 5. Visual representation of float service 6 m from the net test.

2.4. Statistical Analysis

The intraclass correlation coefficient (ICC) was used on the subsample group with the same age to measure test-retest reliability. According to Koo and Li [30], the ICC was considered poor if less than 0.39, between 0.40 and 0.59 was fair, from 0.60 to 0.75 was good reliability, and considered excellent if larger than 0.75. Absolute reliability was analyzed to determine the usefulness of the specific kinesthetic differentiation tests by calculating the standard error of measurement (SEM) using the formula: [$SEM = SD \times \sqrt{1 - ICC}$]. The smallest worthwhile change (SWC) was assumed by multiplying the between-participant standard deviation with different effect sizes (0.2, 0.6, and 1.2). The usefulness can be rated as "good", "marginal", and "satisfactory" when the SEM is below, similar, or higher than the SWC, respectively [31]. Furthermore, minimal detectable change (MDC) was analyzed with the following formula: [$MDC = SEM \times 1.96 \times \sqrt{2}$] to monitor progress so that intra-trial variations do not inaccurately suggest a change. The normality of the sample was tested using the Kolmogorov–Smirnov test ($p > 0.05$). The applied tests were assessed for their ability to distinguish differences between players in different playing positions and situational performance differences of young volleyball players in terms of discriminative ability. Therefore, one-way ANOVA was analyzed to determine differences in five positions (setter vs. passer-hitter vs. opposite player vs. middle blocker vs. libero), while the Student's paired t-test was used to determine if there is a difference between situational performance (successful vs. less successful). Additionally, Cohen's d was calculated with the magnitude of d was qualitatively interpreted using the following thresholds: <0.2, trivial; 0.2 to 0.6, small; 0.6 to 1.2, moderate; 1.2 to 2.0, large; and 2.0 to 4.0, very large [32]. The statistical analysis was performed using SPSS Statistics 27.0 for Windows. The statistical significance for all tests was set at $p < 0.05$.

3. Results

The relative and absolute reliability of specific kinesthetic differentiation ability tests is presented in Table 2. The values of ICC range from 0.66 to 0.87, indicating excellent test-retest reliability in all tests except in the FSTnet, whose reliability was rated as good. The SEM values for almost all of the test measurements were low except in the FSTnet. Furthermore, SEM values were higher than SWC for all variables except for FST6m when the smallest level (0.2 multiplied by between-participants SD) was considered.

Table 2. Results of the reliability and usefulness of the coordination tests.

	Test	Re-Test	ICC	SEM	SWC (0.2, 0.6, 1.2)	MDC
FPT (cm)	62.6 ± 28.7	51.5 ± 16.8	0.84	0.06	0.03, 0.09, 0.19	0.16
OPT (cm)	24.2 ± 11.1	34.8 ± 21.1	0.78	0.07	0.03, 0.09, 0.18	0.19
FSTnet (cm)	111.8 ± 64.6	110.2 ± 46.7	0.66	0.15	0.08, 0.24, 0.48	0.41
FSTx (cm)	59.6 ± 14	54.2 ± 21.7	0.77	0.08	0.03, 0.09, 0.19	0.22
FST6m (cm)	61.4 ± 34.1	77.5 ± 21.6	0.87	0.01	0.04, 0.13, 0.27	0.03

Legend: All values are presented as mean ± SD; ICC = intraclass correlation coefficient; SEM = standard error of measurement; SWC = smallest worthwhile change; MDC = minimal detectable change.

In Table 3, inter-positional differences were presented by applying ANOVA on the whole sample of female volleyball players. The results showed that there were no significant differences in all five specific tests ($p > 0.05$).

Table 3. Inter-positional differences analyzed by one-way ANOVA for the specific kinesthetic differentiation tests on the total sample of 98 young female volleyball players.

Variables	Setter (n = 17)	Passer-Hitter (n = 35)	Opposite Player (n = 16)	Middle Blocker (n = 19)	Libero (n = 11)	F	p
FPT (cm)	72.3 ± 32.9	57.6 ± 29	53.5 ± 25.6	59.4 ± 31.6	48.2 ± 25.3	0.51	0.76
OPT (cm)	40.1 ± 12	38.7 ± 16.8	44 ± 14.2	45.1 ± 18.4	36.5 ± 16.6	0.72	0.6
FSTnet (cm)	124 ± 97.2	109.6 ± 60.4	91.7 ± 37.9	154.3 ± 80.4	109.6 ± 21.6	2.27	0.06
FSTx (cm)	65 ± 21.5	72.2 ± 22.6	75.3 ± 37.1	77.8 ± 38.8	75.9 ± 21.6	0.41	0.84
FST6m (cm)	125.1 ± 67.8	119.6 ± 83.3	106.4 ± 63.4	175.8 ± 87.1	110.2 ± 64.2	1.51	0.2

Legend: All values are presented as mean ± SD; F—f test; p—probability value.

Table 4 shows the difference between less successful and more successful female volleyball players. Based on Table 4, successful young female volleyball players achieved better than less successful players in all five-specific kinesthetic-differentiation tests ($p < 0.01$).

Table 4. Differences analyzed by paired t-test for the specific kinesthetic differentiation between less successful and more successful young female volleyball players.

Variables	Less Successful (n = 59)	More Successful (n = 39)	t-Value	p	ES	
					d	95% CI
FPT (cm)	72 ± 31.5	50 ± 24.1	3.92	<0.01	0.76	0.34–1.18
OPT (cm)	51.5 ± 18.6	35.4 ± 11.4	4.80	<0.01	1	0.56–1.42
FSTnet (cm)	154.2 ± 79.2	82.6 ± 24.5	5.44	<0.01	1.13	0.69–1.55
FSTx (cm)	93.9 ± 28.7	58.9 ± 17.5	7.08	<0.01	1.41	0.95–1.84
FST6m (cm)	178.1 ± 74.3	87.9 ± 54.6	6.06	<0.01	1.34	0.89–1.78

Legend: All values are presented as mean ± SD; t—t test; p—probability value; ES—effect size; d—Cohen d; 95% CI—95% confidence interval.

4. Discussion

This study aimed to analyze the reliability, discriminative ability, and usefulness of new specific kinesthetic differentiation battery tests in young female volleyball players. Almost all tests had excellent test-retest reliability, except the FSTnet, whose reliability was good. These results agree with the findings of [33], who tested junior volleyball players in spiking, setting, serving, and passing techniques. Furthermore, performances of service change through different age groups and levels, as well as the accuracy of trajectory, which is critical for effectiveness in young age groups. In such a way, the intention for using three different service tests was to observe which were most suitable for use in this age group. Observing the results of three service tests, it can be seen that the FSTnet has the lowest result in reliability (0.66). Primarily, the biomechanical process of serving must be on a higher level to achieve correct consecutive repetitions. For players to successfully perform the service from a long distance and to have an accuracy of performance, they should have developed coordination along the kinetic chain [34]. Therefore, the service movement starts with placing the foot on the floor during a step; thus, the reaction of the floor occurs, which passes through the entire body. Moreover, young players are limited with the strength of lower and upper muscles, which leads to greater oscillations in the trials of this test. Accordingly, FST6m is the most convenient for testing reliability in young female volleyball players. Neither does the long distance exists between the player and the net nor does strength come to such an extent. The net placed at a greater distance from the service point represents a psychological barrier that pushes players to apply greater force, which results in greater oscillations. Furthermore, when youth players are serving 9 m away from the net, the target is close to the net on the other side, requiring the ball to go over the net in a higher arc while also using considerable force. This is a challenging combination that could present a problem even for more experienced and technically skilled players.

Upon analyzing the usefulness of the tests, it can be observed that the amount of measurement error and noise in the tests ranges from 0.1 to 0.15 m (1% to 32%), indicating that below 10% only differentiate in FPT and FST6m. Other tests, apart from FST6m, whose usefulness was rated as "good," had small performance changes SWC (0.2) and showed "satisfactory" usefulness. Moreover, moderate variations could be detected as the SEM values were lower than SWC (0.6, 1.2) for all tests indicating "good" usefulness. Therefore, all kinesthetic differentiation tests could detect a real change that exceeds 0.6 and 1.2 times the standard test deviation [31]. Additionally, by observing the MDC (the real changes in measured performance from test and retest), values varied from 16, 19, 41, 22, and 0.03 cm. High values of FSTnet (0.41) may raise concerns regarding the precision of the measure. Consequently, all these tests have small targets, which influence the accuracy and lead to higher performance variations [35]. Specifically, other interactions may also influence testing consistency every time players are tested, such as fatigue, stress, pressure, and concentration [36].

For this study, the intention was to establish how well-applied tests will discriminate against young athletes. Therefore, the present study analyzed the discriminative ability with two objectives: (i) to evaluate are differences between playing positions; (ii) to evaluate differences between more successful and less successful young volleyball players in specific kinesthetic-differentiation battery tests. No significant inter-positional differences were observed in all five tests. Volleyball is a late-specialization sport; therefore, individual player position specialization does not begin until the age of 14–15. The young volleyball players in this study have only begun the specialization process, so the position-specific training has not yet impacted the specific development of certain abilities and skills. It is important to master a high level of general coordination abilities afterward by entering the specialization of playing positions to master complex volleyball techniques later on [12,22,28,37]. In addition, volleyball is a skill-based sport where complex performance techniques, both with and without the ball, are characteristic of all player roles. Therefore, it can be assumed that a high level of kinesthetic differentiation is equally necessary for successful play in all player positions in volleyball. Consequently, a significant difference was observed between less successful and more successful volleyball players for all tests. The results are consistent with the conclusions drawn by Pion et al. [28], who highlighted that differences between elite and sub-elite levels of play in volleyball are influenced by motor coordination. Motor coordination could play a crucial role in differentiating those who reach an expert level in female volleyball from those who do not. As such, the results provide support for the notion that overall motor coordination may serve as a useful predictor of an athlete's potential for advancement in skill-based sports such as volleyball. Moreover, successful talent-identified junior volleyball players can be distinguished from unsuccessful ones based on their skill test results, specifically passing and serving technique evaluations by coaches, but not based on their physical characteristics [38].

These results highlight the significant importance of kinesthetic-differentiation ability in volleyball. This ability consists of three components: force (tension) as the strength of the movement, spatial (space) as the angle of joints, and temporal (time) through movement speed, which all contribute to more efficient volleyball techniques [36,39,40]. However, due to its complexity in defining learning methods and the relatively short contact time with the ball, it is essential to focus more on the development of specific kinesthetic-differentiation abilities in young volleyball players [20]. The accuracy of overhead and forearm passing is important for receiving serves, defending the court (especially for precisely playing balls that the opposing team did not hit with a strong spike but played lightly—which is common in younger age groups), as well as for setting up for a spike. The precision of serves enables serving into empty spaces that opposing serve receivers did not cover or targeting players who are known for poor serve reception.

This study has some limitations. Firstly, the sample only consisted of female athletes; future studies should include male athletes. Additionally, the reliability of the tests was established in a non-fatigued state, so further investigations are needed to assess their effectiveness when athletes are fatigued. Although the sample size was not small, it should be even larger when examining differences between playing positions. Lastly, only discriminant ability was analyzed in this study, and future investigations should establish convergent validity by comparing the results with other tests considered the gold standard.

5. Conclusions

The findings of this study confirm the importance of kinesthetic differentiation ability in volleyball and the need for its development in young athletes. The newly constructed tests of specific kinesthetic differentiation in this study are reliable. Due to their discriminative ability among performance levels, they can be used in the talent identification process and to assess progress in individual player positions in volleyball. Coaches should be aware of the importance of kinesthetic-differentiation ability in complex tasks such as service tests, overhead passing, and forearm passing and incorporate appropriate training methods from an early age. It can be assumed that the FSTx test is suitable for even younger volleyball players than those in this study group, while the FST6m test is best for this group, and it can be assumed that the FSTnet test will be the best option for a cadet, junior, and senior players (U17 and older age groups).

Author Contributions: Conceptualization, K.Đ. and Z.G.; methodology, K.Đ. and G.K.; formal analysis, K.Đ. and G.K.; investigation, K.Đ.; data curation, K.Đ. and Z.G.; writing—original draft preparation, K.Đ.; writing—review and editing, Z.G. and G.K.; supervision, Z.G. All authors have read and agreed to the published version of the manuscript.

Funding: This study received no external funding.

Institutional Review Board Statement: The study was conducted in accordance with the 1975 Declaration of Helsinki, and ethical approval was given by the Ethical Committee of the University of Split (number: 2181-205- 02-05-20-025; 3 December 2020).

Informed Consent Statement: Informed consent was obtained from all subjects involved in the study.

Data Availability Statement: Data can be provided on reasonable request.

Acknowledgments: The authors wish to thank all players and coaches for their contributions to this study.

Conflicts of Interest: The authors declare no conflict of interest.

References

1. Grgantov, Z.; Katić, R.; Janković, V. Morphological Characteristics, Technical and Situation Efficacy of Young Female Volleyball Players. *Coll. Antropol.* **2006**, *30*, 87–96. [PubMed]
2. Milić, M.; Grgantov, Z.; Chamari, K.; Ardigò, L.P.; Bianco, A.; Padulo, J. Anthropometric and Physical Characteristics Allow Differentiation of Young Female Volleyball Players According to Playing Position and Level of Expertise. *Biol. Sport* **2017**, *34*, 19–26. [CrossRef]
3. Sheppard, J.M.; Cronin, J.B.; Gabbett, T.J.; Mcguigan, M.R.; Etxebarria, N.; Newton, R.U. Relative Importance of Strength, Power, and Anthropometric Measures to Jump Performance of Elite Volleyball Players. *J. Strength Cond. Res.* **2008**, *22*, 758–765. [CrossRef] [PubMed]
4. Boichuk, R.; Iermakov, S.; Kovtsun, V.; Pasichnyk, V.; Melnyk, V.; Lazarenko, M.; Kovtsun, V. Effect of Physical Development Parameters and Conditioning Abilities on the Level of Motor Coordination in Female Volleyball Players in the Phase of Specialized Basic Training. *J. Phys. Educ. Sport* **2018**, *18*, 1950–1957.
5. Šimonek, J. The Effect of Intervention on the Changes of Coordination Factors in the Youth Sports Preparation. *Sport Sci.* **2016**, *9*, 77–81.
6. Ford, P.; de Ste Croix, M.; Lloyd, R.; Meyers, R.; Moosavi, M.; Oliver, J.; Till, K.; Williams, C. The Long-Term Athlete Development Model: Physiological Evidence and Application. *J. Sports Sci.* **2011**, *29*, 389–402. [CrossRef] [PubMed]
7. Hirtz, P.; Starosta, W. Sensitive and Critical Periods of Motor Co-Ordination Development and Its Relation to Motor Learning. *J. Hum. Kinet.* **2002**, *7*, 19–28.
8. Viru, A.; Loko, J.; Harro, M.; Volver, A.; Laaneots, L.; Viru, M. Critical Periods in the Development of Performance Capacity During Childhood and Adolescence. *Eur. J. Phys. Educ.* **1999**, *4*, 75–119. [CrossRef]
9. Van Hooren, B.; De Ste Croix, M. Sensitive Periods to Train General Motor Abilities in Children and Adolescents: Do They Exist? A Critical Appraisal. *Strength Cond. J.* **2020**, *42*, 7–14. [CrossRef]
10. Quatman-Yates, C.C.; Quatman, C.E.; Meszaros, A.J.; Paterno, M.V.; Hewett, T.E. A Systematic Review of Sensorimotor Function during Adolescence: A Developmental Stage of Increased Motor Awkwardness? *Br. J. Sports Med.* **2012**, *46*, 649–655. [CrossRef]
11. Lech, G.; Jaworski, J.; Lyakh, V.; Krawczyk, R. Effect of the Level of Coordinated Motor Abilities on Performance in Junior Judokas. *J. Hum. Kinet.* **2011**, *30*, 153–160. [CrossRef] [PubMed]
12. Malikova, A.N.; Doroshenko, E.Y.; Symonik, A.V.; Tsarenko, E.V.; Veritov, A.I. The Ways of Improvement Special Physical Training of High-Qualified Women Volleyball Players in Competitive Period of Annual Macrocycle. *Phys. Educ. Stud.* **2018**, *22*, 38. [CrossRef]
13. Kamandulis, S.; Venckunas, T.; Masiulis, N.; Matulaitis, K.; Balčiunas, M.; Peters, D.; Skurvydas, A. Relationship between General and Specific Coordination in 8- to 17-Year-Old Male Basketball Players. *Percept. Mot. Ski.* **2013**, *117*, 821–836. [CrossRef]
14. Trecroci, A.; Cavaggioni, L.; Caccia, R.; Alberti, G. Jump Rope Training: Balance and Motor Coordination in Preadolescent Soccer Players. *J. Sports Sci. Med.* **2015**, *14*, 792–798. [PubMed]
15. Šimonek, J. *Coordination Abilities in Volleyball;* De Gruyter Open: Warsaw, Poland, 2014.
16. Stojanović, N.; Stojanović, T.; Stojanović, D.; Herodek, K.; Jurko, D. The Influence of Coordination Abilities on the Precision of Forearm Passing in Volleyball. *Defendologija* **2014**, *1*, 75–83. [CrossRef]
17. Bańkosz, Z.; Stefaniak, T. Elbow Joint Position and Hand Pressure Force Sense under Conditions of Quick Reaction in Table Tennis Players. *Kinesiology* **2021**, *53*, 95–103. [CrossRef]
18. Boichuk, R.; Iermakov, S.; Korop, M.; Kovtsun, V.; Vaskan, I.; Shankovskyi, A.; Kovtsun, V. Coordination Training of 16–17-Year-Old Volleyball Players (Girls). *J. Phys. Educ. Sport* **2020**, *20*, 2976–2983.
19. Brod'áni, J.; Šimonek, J. Prediction of Coordination Performance in Ice-Hockey Players Based on the Structure of Coordination Capacities. *Palestrica Third Millenn. Civiliz. Sport* **2012**, *13*, 316–320.
20. Szczepan, S.; Wróblewska, Z.; Klarowicz, A.; Błacha, R.; Rejman, M. Effects of Different Forms of Extrinsic Feedback on the Accuracy of Force Production and to Differentiate This Force in the Simple Cyclic Movements of the Upper and Lower Limb. *Pol. Hyperb. Res.* **2020**, *72*, 39–56. [CrossRef]
21. Brod'áni, J.; Šimonek, J. Structure of Coordination Capacities and Prediction of Coordination Performance in Sport Games. *Stud. Univ. Babes Bolyai Educ. Artis Gymnast.* **2010**, *55*, 3–10.
22. Đolo, K.; Grgantov, Z.; Milić, M. Absolute and Relative Reliability of Tests Assessing Kinesthetic Differentiation Ability in Youth Female Volleyball Players. *Acta Kinesiol.* **2019**, *13*, 44–49.
23. Kozina, Z.; Venckūnas, T.; Masiulis, N.; Matulaitis, K.; Balčiūnas, M.; Peters, D.; Skurvydas, A. The Influence of a Special Technique for Developing Coordination Abilities on the Level of Technical Preparedness and Development of Psycho-Physiological Functions of Young Volleyball Players 14–16 Years of Age. *J. Phys. Educ. Sport* **2018**, *18*, 1445–1454.
24. Boichuk, R.; Iermakov, S.; Kovtsun, V. Special Aspects of Female Volleyball Players' Coordination Training at the Stage of Specialized Preparation. *J. Phys. Educ. Sports* **2017**, *17*, 884–891.
25. Uljevic, O.; Esco, M.R.; Sekulic, D. Reliability, Validity, and Applicability of Isolated and Combined Sport-Specific Tests of Conditioning Capacities in Top-Level Junior Water Polo Athletes. *J. Strength Cond. Res.* **2014**, *28*, 1595–1605. [CrossRef]
26. Gabbett, T.; Georgieff, B.; Anderson, S.; Cotton, B.; Savovic, D.; Nicholson, L. Changes in Skill and Physical Fitness Following Training in Talent-Identified Volleyball Players. *J. Strength Cond. Res.* **2006**, *20*, 29–35. [CrossRef]

27. Rikberg, A.; Raudsepp, L. Multidimensional Performance Characteristics in Talented Male Youth Volleyball Players. *Pediatr. Exerc. Sci.* **2011**, *23*, 537–548. [CrossRef]
28. Pion, J.A.; Fransen, J.; Deprez, D.N.; Segers, V.I.; Vaeyens, R.; Philippaerts, R.M.; Lenoir, M. Stature and Jumping Height Are Required in Female Volleyball, but Motor Coordination Is a Key Factor for Future Elite Success. *J. Strength Cond. Res.* **2015**, *29*, 1480–1485. [CrossRef] [PubMed]
29. Drust, B.; Waterhouse, J.; Atkinson, G.; Edwards, B.; Reilly, T. Circadian Rhythms in Sports Performance—An Update. *Chronobiol. Int.* **2005**, *22*, 21–44. [CrossRef]
30. Koo, T.K.; Li, M.Y. A Guideline of Selecting and Reporting Intraclass Correlation Coefficients for Reliability Research. *J. Chiropr. Med.* **2016**, *15*, 155–163. [CrossRef]
31. Hopkins, W. How to Interpret Changes in an Athletic Performance Test. *Sportscience* **2004**, *8*, 1–17.
32. Hopkins, W.G. Measures of Reliability in Sports Medicine and Science. *Sport. Med.* **2000**, *30*, 1–15. [CrossRef]
33. Gabbett, T.J.; Georgieff, B. The Development of a Standardized Skill Assessment for Junior Volleyball Players. *Int. J. Sports Physiol. Perform.* **2006**, *1*, 95–107. [CrossRef]
34. Bieleke, M.; Kriech, C.; Wolff, W. Served Well? A Pilot Field Study on the Effects of Conveying Self-Control Strategies on Volleyball Service Performance. *Behav. Sci.* **2019**, *9*, 93. [CrossRef]
35. Bańkosz, Z. The Kinesthetic Differentiation Ability of Table Tennis Players. *Hum. Mov.* **2012**, *13*, 16–21. [CrossRef]
36. Struzik, A.; Rokita, A.; Pietraszewski, B.; Popowczak, M. Accuracy of Replicating Static Torque and Its Effect on Shooting Accuracy in Young Basketball Players. *Hum. Mov.* **2014**, *15*, 216–220. [CrossRef]
37. Mostaert, M.; Pion, J.; Lenoir, M.; Vansteenkiste, P. A Retrospective Analysis of the National Youth Teams in Volleyball: Were They Always Faster, Taller, and Stronger? *J. Strength Cond. Res.* **2022**, *36*, 2615–2621. [CrossRef] [PubMed]
38. Gabbett, T.; Georgieff, B.; Domrow, N. The Use of Physiological, Anthropometric, and Skill Data to Predict Selection in a Talent-Identified Junior Volleyball Squad. *J. Sports Sci.* **2007**, *25*, 1337–1344. [CrossRef]
39. Pakosz, P. EMG Parameters and Kinesthetic Differentiation during the Free-Throw of Basketball Players with Various Levels of Athletic Experience. *Cent. Eur. J. Sport Sci. Med.* **2013**, *2*, 31–38.
40. Moradi, M.; Ghorbani, A.; Yazdanpanah, M.; Javan Eghbal Tajeddin, G.; Bahrami, A. Relationship between Trait and State Anxiety with Force Control and Accommodation of Dominant Hand's Angle in Male Students. *Ann. Appl. Sport Sci.* **2015**, *3*, 29–38. [CrossRef]

Journal of
Functional Morphology and Kinesiology

MDPI

Article

Countermovement, Hurdle, and Box Jumps: Data-Driven Exercise Selection

M. Tino Janikov [1], Jan Pádecký [1], Valentin Doguet [2] and James J. Tufano [1,*]

1 Sport Sciences-Biomedical Department, Faculty of Physical Education and Sport, Charles University, 162 52 Prague, Czech Republic; tino.janikov@gmail.com (M.T.J.)
2 Atelier Maker®, 497 00 Doué-en-Anjou, France
* Correspondence: tufano@ftvs.cuni.cz; Tel.: +420-774-367-235

Abstract: Apart from squat jumps, countermovement jumps (CMJ), and drop jumps, differences among other jump variations are not as well researched, making data-driven exercise selection difficult. To address this gap, this study compared selected concentric and eccentric jump parameters of maximal effort CMJ, hurdle jumps over 50 cm hurdle (HJ), and box jumps onto a 50 cm box (BJ). Twenty recreationally trained men (25.2 ± 3.5 years) performed 3 repetitions of CMJs, HJs, and BJs, each on separate days. The data were collected using force platforms and a linear position transducer. The mean of 3 trials of each jump variation was analyzed using repeated measures ANOVA and Cohen's d. Countermovement depth was significantly greater ($p \leq 0.05$) and peak horizontal force significantly lower during CMJ compared to HJ and BJ. However, there were no differences in peak velocity, peak vertical and resultant force, and total impulsion time. Finally, BJ significantly decreased peak impact force by ~51% compared to CMJ and HJ. Therefore, the propulsive parameters of HJ and BJ seem to be similar to CMJ, despite CMJ having a greater countermovement depth. Furthermore, overall training load can be decreased dramatically by using BJ, which reduced peak impact force by approximately half.

Keywords: plyometric training; stretch-shortening cycle; exercise variation

Citation: Janikov, M.T.; Pádecký, J.; Doguet, V.; Tufano, J.J. Countermovement, Hurdle, and Box Jumps: Data-Driven Exercise Selection. *J. Funct. Morphol. Kinesiol.* **2023**, *8*, 61. https://doi.org/10.3390/jfmk8020061

Academic Editor: Diego Minciacchi

Received: 31 March 2023
Revised: 5 May 2023
Accepted: 8 May 2023
Published: 11 May 2023

1. Introduction

Plyometric training is widely used in strength and conditioning practice to increase power output [1]. The cornerstone of plyometric training involves movements that maximize the use of the stretch shortening cycle, which is defined as eccentric pre-stretch directly followed by an isometric amortization phase and explosive concentric muscle action [2] that results in augmented power expression compared to purely concentric explosive movements [3]. Although bodyweight countermovement jumps (CMJ) are commonly used lower body plyometric exercises [4,5], other variations, such as hurdle jumps (HJ) and box jumps (BJ), are also performed. Although these variations are often used interchangeably in practice, the nature of overcoming an obstacle like a hurdle or a box might result in changes to concentric and eccentric jump parameters, which may affect the resultant training adaptations over time. However, these jump types have not yet been compared in a single study; therefore, the magnitude of the differences are still unknown.

To the best of our knowledge, only two studies [6,7] have previously investigated these differences, but it is not yet possible to draw practical conclusions from their findings due to the following methodological shortcomings. The first study compared a single CMJ performed from a static standing position to the 3rd out of 4 continuous HJs over 4 hurdles [6]. The static nature of the CMJ and dynamic nature of the continuous HJs makes it difficult to attribute any differences between them to the specific jump type. Therefore, it is necessary to compare CMJ with HJ, where both variations are performed either as a single jump or as a sequence of continuous jumps. The second study reported no differences in propulsive parameters between jumps onto boxes of two different heights, as well as no relationship between maximal CMJ height and maximal achievable BJ height [7]. However, this study did not compare any other parameters between CMJ and BJ, leaving the differences between the jump variations largely unexplored.

Although some of the differences among CMJ, HJ, and BJ may seem intuitive, such as less impact force during BJ due to the lack of gravitational acceleration after reaching the apex of the jump [8], quantifying the magnitudes of these difference is important. As the exercise selection represents an important load management tool for coaches aiming to acutely decrease eccentric load (i.e., basketball, volleyball, etc.) while continuing structured concentric training, it would be desirable to base training prescriptions on data rather than solely on intuition.

Furthermore, some changes in jump technique may be present when jumping over or atop an obstacle compared to purely vertical jumping due to different direction and magnitudes of concentric forces as a result of horizontal movement [9,10]. Additionally, humans are likely to adjust their performance based on expected demands of the task at hand [11–14] (i.e., the expected impact upon landing or perceived fear of hitting the obstacle mid-flight), which might lead to differences in propulsive jump phases between the jump variations. Therefore, the purpose of the present study was to assess and quantify the differences among CMJ, HJ, and BJ in the same study in order to help guide coaching decisions. Based on the aforementioned points, we hypothesized that the BJ would result in less impact force than the other two variations. Furthermore, we hypothesize that subjects would jump with a greater countermovement depth in order to maximize the propulsion time during the HJ and BJ, thereby leading to greater forces and velocities in order to overcome the hurdle and the box when compared to the traditional CMJ.

2. Materials and Methods

2.1. Participants

Twenty recreationally trained university-aged men volunteered to participate in this study (25.2 ± 3.5 years, 180.2 ± 4.4 cm, 80.0 ± 7.8 kg, 11.5 ± 2.7% body fat, CMJ height 49.5 ± 6.4 cm). This sample size was shown to be appropriate by a-priori power analysis (effect size = 0.8; α err. prob. = 0.05; Power (1–β err. prob.) = 0.90) using G*Power 3.1.9.7 (RRID:SCR_013726), resulting in a necessary sample size of 19 subjects. The subjects had experience in sports where jumping was common, such as soccer, basketball, handball, track and field, and martial arts, but none of the subjects competed in any of these sports professionally. All subjects were able to perform countermovement vertical jumps and hold a light wooden dowel racked across the posterior shoulders without pain. The subjects reported no ongoing rehabilitation process post-injury or any other chronic conditions that could influence the results or prevent them from safely participating in this study. The subjects were asked to arrive rested (≥7 h of sleep and no exhausting lower body training 36 h before testing), well hydrated, and having fasted (≥2 h). Additionally, we asked the subjects to maintain their habitual dietary and supplementation intake while participating in the study. This study was completed in accordance with the Declaration of Helsinki. The experimental procedures were approved by the university Ethics Board (126/2018) and were explained to the subjects prior to providing institutionally approved written informed consent to participate in the study.

2.2. Data Collection

A quasi-randomized experimental approach was used to quantify the differences between the CMJ, HJ, and BJ in concentric and eccentric jump parameters. The experimental design is depicted in Figure 1. Every subject completed 3 laboratory visits separated by at least 48 h. All visits were scheduled at the same time of day (±1 h). The first visit began with measurements of body weight, body height, and body composition performed using an electronic column scale with a fitted stadiometer (Seca 769, Seca 220; Seca Ltd., Hamburg, Germany) and bioelectric impedance (InBody 720; Biospace Co., Ltd., Seoul, Republic of Korea), respectively. The subjects then completed a standardized warm up which included a single set per exercise of in-place running (30 ground contacts per leg), 10 in-place bilateral pogo jumps, dynamic unilateral stretches of hip, knee, and ankle muscles (3 repetitions per leg, per exercise), 10 bodyweight squats, followed by reverse lunges, side lunges, unilateral stiff-legged deadlifts, and supine lying unilateral hip raises for 5 repetitions without added load per leg for each exercise.

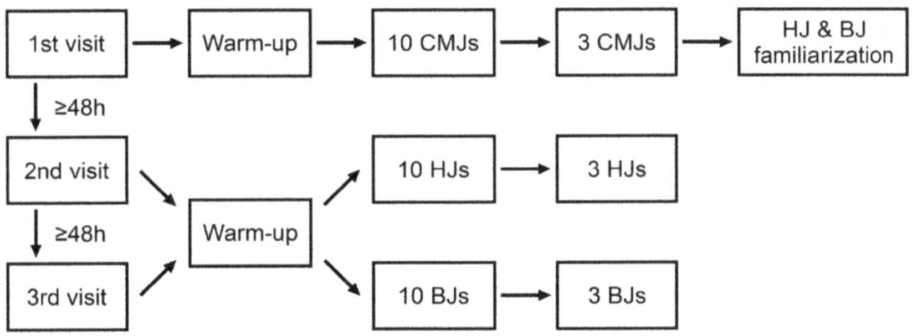

Figure 1. Experimental design.

Next, to provide a specific warm-up and remove any possible effect of potentiation that could positively or negatively influence the experimental jumps, 10 preparatory CMJs without arm swing were performed, with the light wooden dowel held across their posterior shoulders behind the base of the neck. These jumps were performed using a running 10-s timer, allowing for roughly 8 s to prepare for the next jump. This jump frequency and volume were selected based on pilot testing, which showed no negative effects on following jumping performance, while allowing for a safe return to the starting position and preparation for the next jump in all 3 jump types. The total time necessary to complete the warm-up process was approximately 5 to 8 min.

After completing the warm-up, the subjects performed 3 maximal CMJs using the same jump technique and 10-s running timer as during the preparatory jumps. The subjects were instructed to jump as high as possible and to land softly while receiving verbal encouragement from the researchers. The countermovement depth, speed, and stance width during all jumps were self-selected by the subjects to maintain ecological validity. To conclude the first visit, the subjects were familiarized with the HJ and BJ procedures by performing a minimum of 3 repetitions of HJ and BJ. Additional repetitions were allowed if requested by a subject or considered necessary by the researchers.

At the beginning of the second and third visits, the subjects performed the same standardized warm-up, followed by 10 warm-up jumps, and 3 maximal jumps over the 50 cm hurdle or onto the 50 cm box with the wooden dowel held across the posterior shoulders as described above. The order of HJ and BJ was randomized. After the subjects jumped over the hurdle or onto the box, they stepped back to the starting position and prepared themselves to perform the next repetition at the end of the 10-s period. A visual depiction of the set-up including the athlete, linear position transducer, force plate(s), box, and hurdle is shown in Figure 2.

The ground reaction force data of all maximal jumps were recorded using two synchronized piezoelectric force plates (Kistler 9286BA; Kistler Instruments Inc., Winterthur, Switzerland). The force plates were placed side-by-side on the ground to measure CMJs. The first force plate, used to measure propulsive forces for both HJ and BJ, was positioned on the ground separated from the hurdle or box by 15 cm. The second force plate was used to measure landing forces. The position of the second force plate was on the ground 5 cm behind the hurdle or on top of the box 5 cm from the edge closest to the subject.

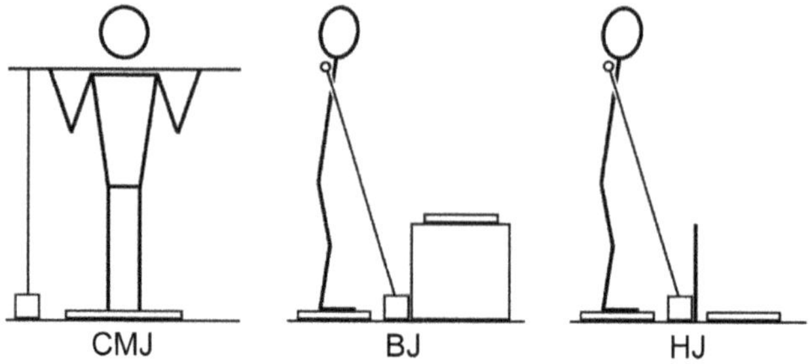

Figure 2. Data collection setup during the countermovement jump (CMJ), box jump (BJ), and hurdle jump (HJ) conditions.

The ground reaction forces were recorded using sampling frequencies of 1000 Hz, a 16-bit A/D board amplifier, and BioWare v5.3.2.9 software (Kistler Instruments Inc., Winterthur, Switzerland). A custom MATLAB program (1.8.0.121; MathWorks, Natic, MA, USA) was used to calculate peak propulsive forces (PF) in vertical, horizontal, and resultant directions as maximal ground reaction force produced in each direction during the propulsive phase, total impulsion time as the time from the beginning of the countermovement to the take-off, and peak impact force as maximal resultant ground reaction force produced during the landing phase. A threshold of 20 N was used to identify individual phases of the jump.

The velocity and displacement data of all maximal jumps were recorded using a linear position transducer (GymAware Power Tool; Kinetic Performance Technology Pty. Ltd., Canberra, Australia). The string of the linear position transducer was attached to the wooden dowel 30 cm from the right shoulder to the end of the dowel [15]. The linear position transducer was placed on the ground directly below the end of the string for CMJ, and in the middle of the horizontal distance between the force plates in line with the attachment of the string for HJ and BJ. The attachment of the string and the position of the linear position transducer was selected due to the technical restrictions of the equipment used and the motion necessary to complete the measured task. However, the linear position transducer used in this study automatically corrects the data for horizontal displacement and allows the three conditions to be compared. The correct position of the string was checked before every jump. Additionally, one researcher observed the movement of the dowel during the data collection. Any trials with notable rotational dowel movements, deviation from horizontal dowel position, or other cases of failed trial (i.e., hitting the obstacle during the flight or not landing on the force plate) were excluded, and the trial was repeated after completing the prescribed rest interval. The internal software was used to calculate peak concentric velocity as maximal velocity of the dowel during the concentric propulsive phase, and countermovement depth as a maximum downward displacement of the dowel below the standing position during the propulsive phase.

2.3. Statistical Analysis

Data acquired during 3 repetitions of each jump type were averaged for each subject and used for analysis. The Shapiro-Wilk test and Quantile-Quantile plots were used to test the data for normality of distribution. Means and standard deviations (SDs) were calculated for all variables and repeated ANOVA measures were performed to assess the data. A Greenhouse-Geisser correction was used in instances where sphericity was not assumed. Pairwise comparisons were performed using a Holm-Bonferroni follow-up when appropriate. Alpha level for significance for all tests was set at $\rho \leq 0.05$. Effect sizes (Cohen's d) were calculated and interpreted as trivial (<0.20), small (0.20 to 0.49), moderate (0.50 to 0.79), and large ($\geq$0.80) [16]. For the sake of clarity and concision, only moderate and large effect sizes will be discussed in further sections. All statistical analyses were performed using RStudio 2022.07.2 + 576 (Integrated Development Environment for R;

RStudio, PBC; Boston, MA, USA). The countermovement depth, peak concentric velocity, and total impulsion time data are presented as absolute values; all ground reaction force data are presented as relative to the subject's bodyweight ($N \cdot N^{-1}$). The intra-day intraclass correlation coefficients were calculated and interpreted as poor (<0.50), moderate (0.50 to 0.74), good (0.75 to 0.90), and excellent (>0.90) reliability [17].

3. Results

All variables have shown good to excellent reliability, with the only exception being moderate reliability of peak impact force during BJ (Figures 3–6). Nevertheless, a total of five trials had to be discarded and repeated (3 CMJs, 1 HJ, 1 BJ) due to excessive dowel movement (4 trials) or missing the force plate upon landing (1 trial). No significant differences existed for PF-vertical, PF-resultant, peak concentric velocity, and total impulsion time (Figures 3 and 4, respectively). However, a non-significant moderate effect was present for longer total impulsion time during CMJ compared to HJ (Figure 4b). There were significant moderate and small effects for deeper countermovement depth during CMJ compared to HJ and BJ, respectively (Figure 5a). Finally, significant large effects were shown for less PF-horizontal during CMJ compared to both HJ and BJ (Figure 5b), and for less peak impact force during BJ compared to both CMJ and HJ (Figure 6).

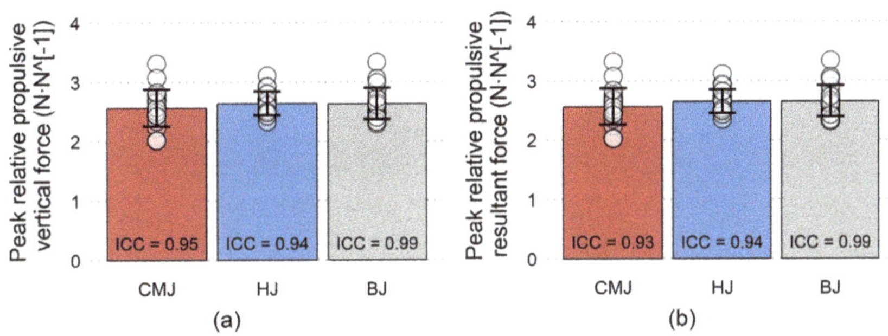

Figure 3. Mean (±SD) and inter-day interclass correlation (ICC) for (**a**) peak vertical propulsive force relative to bodyweight and (**b**) peak resultant propulsive force relative to bodyweight during countermovement jump (CMJ), hurdle jump (HJ), and box jump (BJ).

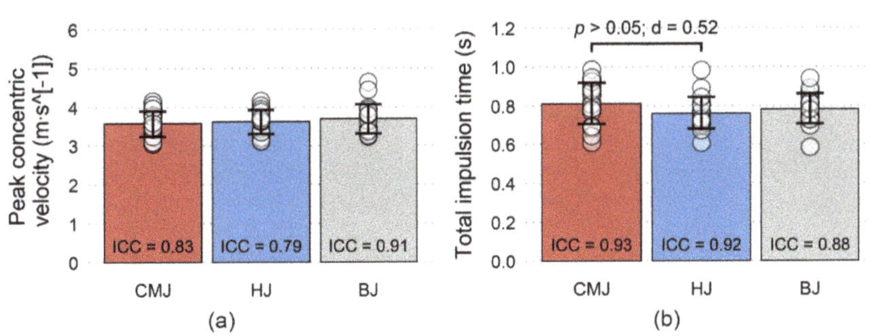

Figure 4. Mean (±SD) and inter-day interclass correlation (ICC) for (**a**) peak concentric velocity and (**b**) total impulsion time during countermovement jump (CMJ), hurdle jump (HJ), and box jump (BJ).

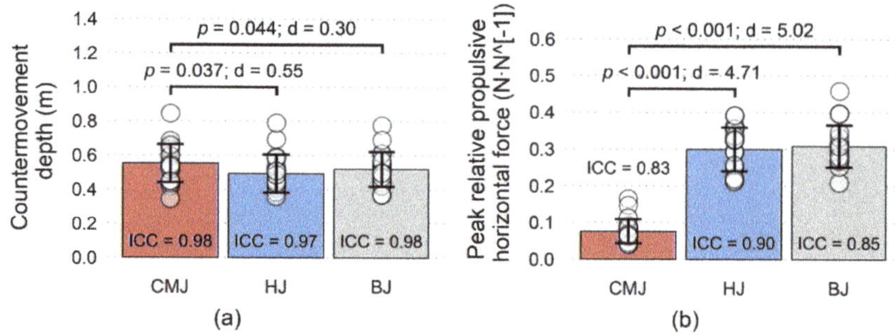

Figure 5. Mean (±SD) and inter-day interclass correlation (ICC) for (**a**) countermovement depth and (**b**) peak horizontal propulsive force relative to bodyweight during countermovement jump (CMJ), hurdle jump (HJ), and box jump (BJ).

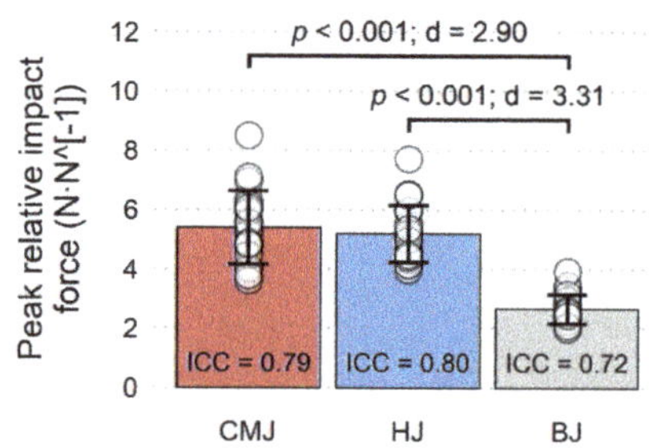

Figure 6. Mean (±SD) and inter-day interclass correlation (ICC) for peak resultant impact force relative to bodyweight during countermovement jump (CMJ), hurdle jump (HJ), and box jump (BJ).

4. Discussion

The results of the present study show that the subjects adjusted the propulsive jump phase in response to the added obstacle during HJ and BJ compared to the CMJ condition. These adjustments were manifested as decreased countermovement depth and increased PF-horizontal. However, there were no significant differences in any of the other propulsive variables (PF-vertical, PF-resultant, peak concentric velocity, and total impulsion time). Furthermore, as expected, BJ significantly reduced peak impact force compared to HJ and CMJ conditions. These results partly support our initial hypothesis, but not fully, as the different jump types did not significantly differ in all measured variables.

This study's most important finding is the lack of significant effects of jump type on PF-vertical, PF-resultant, and peak concentric velocity. Even though HJ and BJ conditions resulted in significantly larger PF-horizontal compared to CMJ, this approximately four-fold PF-horizontal difference was not sufficient to significantly influence the PF-vertical and PF-resultant. Similarly, a previous study reported non-significant trivial-to-small effects of box height for peak propulsive force, peak propulsive power, propulsive rate of force development, and concentric time to take-off [7]. This supports the possibility of interchangeably using CMJ, HJ, and BJ (from a propulsive perspective) in training practice, as similar magnitudes of propulsive forces and velocities likely impose similar training stimulus.

To maintain ecological validity, the subjects were allowed to adopt their preferred countermovement depth during each condition, which resulted in an interesting and unexpected finding. Previous studies have shown a relationship between countermovement depth and total impulsion time [18–20], and altering total impulsion time by manipulating countermovement depth can yield two different acute benefits. Increasing countermovement depth can allow for greater concentric work to be produced via greater available distance, which can also yield higher velocities [18–22]. Such exercise variations would find their place as sport-specific stimulus in athletes performing explosive motions from deep squat positions (i.e., ski jumps, Olympic weightlifting, swimming, sumo, etc.). On the other hand, decreasing the countermovement depth (i.e., decreasing the range of motion) and therefore decreasing the total impulsion time might allow for the more efficient utilization of the stretch-shortening cycle by increasing the rate of force development and power production in both eccentric and concentric jump phases [23], as well as increased eccentric work, amortization, and concentric force, and decreasing amortization time [24]. Therefore, exercise variations utilizing quicker jumps with smaller countermovement depth would be more specific in athletes taking advantage of the stretch-shortening cycle (i.e., basketball, volleyball, high jump, gymnastics, etc.).

Contrary to previous research, our results show significantly greater countermovement depth during CMJ compared to the other two included jump types, but this difference did not manifest in a significant difference in total impulsion time, PF-vertical, PF-resultant, or peak concentric velocity. The only moderate effect of jump type observed for these propulsive variables was a non-significantly shorter total impulsion time during HJ compared to CMJ. This discrepancy in comparison to the previous research might be a result of countermovement depth being measured as a downward shoulder displacement in our study. Therefore, smaller countermovement depth during HJ and BJ might be indicative of subjects keeping a slightly more upright torso position with the presence of the obstacle. However, the adjustments resulting from the added obstacle and leading to differences in countermovement depth and PF-horizontal were not large enough to meaningfully change the other dependent propulsive variables. Based on this, we consider HJ and BJ to be valid alternative variations to CMJ from the perspective of measured acute propulsive parameters.

As expected, there was a significant large effect of BJ for reducing peak impact force compared to HJ and CMJ. Our results are in line with previous research showing a significantly reduced sum of ankle, knee, and hip joint peak landing power during BJ compared to CMJ and HJ [8]. Nevertheless, the magnitude of peak impact force reduction during the BJ (~51%) was quite remarkable. Considering this, the main factors causing the reduction of peak impact force were probably a combination of instructions for a soft landing and a box height (50 cm) that coincidentally matched the mean jump height of the subjects (49.5 cm), causing decreased time for downward acceleration and subsequent lower landing velocity. Therefore, future research might take this a step further and evaluate the differences in landing forces during box-to-CMJ-height ratios other than the nearly 1:1 ratio that was present in this study.

Even though improving eccentric strength and landing mechanics are warranted in certain situations [25], other situations (i.e., periods of increased training and/or competitive load, or acute patellofemoral sensitivity) might require lowering impact force for load management [26,27]. For example, acute patellofemoral pain seems to decrease an athlete's ability to effectively absorb impact force [28]. In turn, increased magnitudes and rates of eccentric forces developed in the patellofemoral joint during landing were significantly correlated with increased patellofemoral pain in young symptomatic women [29]. Therefore, according to our data, coaches can effectively use BJ to reduce peak impact force by ~51% compared to CMJ, while not decreasing concentric performance when decreased eccentric loading is warranted.

Although the present study provided data for evidence-based decision-making when prescribing three plyometric exercises, much is still unknown regarding this topic. For example, the effect of an enhanced stretch-shortening cycle by performing multiple continuous repetitions, fatigue resulting from a higher volume set, the effect of various box heights on IF, and kinematic analysis of the hip, knee, and ankle joint movements during the jump were not assessed in this study. Furthermore, this study had some limitations. Firstly, the attachment of the linear position transducer to one end of the dowel held by the subjects across the shoulders could potentially lead to some error of measurement if the subject's shoulders tilted during the execution of the test. To mitigate the occurrence of this error, the movement of the dowel was observed by one researcher, and the trials where any notable rotational movements or deviation from horizontal position occurred were excluded. Secondly, jump technique (i.e., the amount of forward lean of the trunk) may to some extent influence the countermovement depth results, as the string of linear position transducer was attached to the dowel held across the shoulders. Therefore, future studies should aim to attach the string of the linear position transducer to a subject's waist. Such attachment would allow them to use the arm swing while performing the jumps, which would be beneficial as the real-life training programs would likely not restrict the arm swing. However, this was not possible in the present study due to the specific technological constraints of equipment used during the HJ variation (i.e., string of the linear positional transducer colliding with the top of the hurdle upon landing). Therefore, we decided to use a string attached via the dowel in all jump types for consistency. Finally, studies investigating vertical jumps usually include jump height comparisons. However, the current study was not meant to assess height, but instead to compare important concentric and eccentric parameters influencing training adaptation. In this sense, jump height as an outcome would not have added much value to the discussion.

5. Conclusions

The present study provides data to support two distinct training considerations. Firstly, coaches can use a BJ with a box height similar to the maximum CMJ height to reduce an athlete's peak impact force by approximately half, which can be highly beneficial during certain training periods where impact forces should be reduced. Secondly, coaches should be aware that CMJ results in a significantly greater countermovement of the shoulders compared to the BJ and HJ, which does not seem to provide any benefits for improving other propulsive parameters as no differences were seen in PF-vertical, PF-resultant, peak concentric velocity, or total impulsion time. Thus, a traditional CMJ may not be the ideal exercise choice for athletes who need to perform jumps quickly with minimal countermovement depth. On the other hand, there are other possible factors that might influence the exercise selection, so further research should be conducted in this area before coaches should definitely choose one jump type over another for propulsive training purposes.

Author Contributions: Conceptualization, M.T.J. and J.J.T.; methodology, M.T.J. and J.J.T.; software, V.D.; formal analysis, M.T.J.; investigation, M.T.J. and J.P.; data curation, M.T.J.; writing—original draft preparation, M.T.J.; writing—review and editing, M.T.J., J.P., V.D. and J.J.T.; visualization, M.T.J.; supervision, J.J.T.; project administration, M.T.J.; and funding acquisition, M.T.J., J.P. and J.J.T. All authors have read and agreed to the published version of the manuscript.

Funding: This research was funded by an internal grant (Cooperatio Program, research area SPOB) and an internal student research support (SVV 260599).

Institutional Review Board Statement: The study was conducted in accordance with the Declaration of Helsinki and approved by the Institutional Ethics Committee of Charles University (126/2018, 25 June 2018).

Informed Consent Statement: Informed consent was obtained from all subjects involved in the study.

Data Availability Statement: Data will be available upon a request sent to the corresponding author of this study.

Acknowledgments: We would like to thank all participants for volunteering their time and energy to help us complete this study.

Conflicts of Interest: The authors declare no conflict of interest. The funders had no role in the design of the study; in the collection, analyses, or interpretation of data; in the writing of the manuscript; or in the decision to publish the results.

References

1. Durell, D.L.; Pujol, T.J.; Barnes, J.T. A Survey of the Scientific Data and Training Methods Utilized by Collegiate Strength and Conditioning Coaches. *J. Strength Cond. Res.* **2003**, *17*, 368–373. [CrossRef]
2. Turner, A.N.; Jeffreys, I. The Stretch-Shortening Cycle: Proposed Mechanisms and Methods for Enhancement. *Strength Cond. J.* **2010**, *32*, 87–99. [CrossRef]
3. Bobbert, M.F.; Gerritsen, K.G.; Litjens, M.C.; Van Soest, A.J. Why Is Countermovement Jump Height Greater than Squat Jump Height? *Med. Sci. Sports Exerc.* **1996**, *28*, 1402–1412. [CrossRef] [PubMed]
4. Markovic, G. Does Plyometric Training Improve Vertical Jump Height? A Meta-Analytical Review. *Br. J. Sports Med.* **2007**, *41*, 349–355. [CrossRef] [PubMed]
5. Markovic, G.; Mikulic, P. Neuro-Musculoskeletal and Performance Adaptations to Lower-Extremity Plyometric Training. *Sports Med.* **2010**, *40*, 859–895. [CrossRef]
6. Cappa, D.F.; Behm, D.G. Training Specificity of Hurdle vs. Countermovement Jump Training. *J. Strength Cond. Res.* **2011**, *25*, 2715–2720. [CrossRef]
7. Koefoed, N.; Dam, S.; Kersting, U.G. Effect of Box Height on Box Jump Performance in Elite Female Handball Players. *J. Strength Cond. Res.* **2022**, *36*, 508–512. [CrossRef]
8. Van Lieshout, K.G.; Anderson, J.G.; Shelburne, K.B.; Davidson, B.S. Intensity Rankings of Plyometric Exercises Using Joint Power Absorption. *Clin. Biomech.* **2014**, *29*, 918–922. [CrossRef]
9. Fukashiro, S.; Besier, T.F.; Barrett, R.; Cochrane, J.; Nagano, A.; Lloyd, D.G. Direction Control in Standing Horizontal and Vertical Jumps. *Int. J. Sport Health Sci.* **2005**, *3*, 272–279. [CrossRef]
10. Meylan, C.M.P.; Nosaka, K.; Green, J.; Cronin, J.B. Temporal and Kinetic Analysis of Unilateral Jumping in the Vertical, Horizontal, and Lateral Directions. *J. Sports Sci.* **2010**, *28*, 545–554. [CrossRef]
11. Ferris, D.P.; Liang, K.; Farley, C.T. Runners Adjust Leg Stiffness for Their First Step on a New Running Surface. *J. Biomech.* **1999**, *32*, 787–794. [CrossRef] [PubMed]
12. García-López, D.; Maroto-Izquierdo, S.; Zarzuela, R.; Martín-santana, E.; Antón, S.; Sedano, S. The Effects of Unknown Additional Eccentric Loading on Bench-Press Kinematics and Muscle Activation in Professional Handball and Rugby Players. *Eur. J. Sport Sci.* **2020**, *20*, 1042–1050. [CrossRef]
13. Heiss, D.G.; Shields, R.K.; Yack, H.J. Anticipatory Control of Vertical Lifting Force and Momentum during the Squat Lift with Expected and Unexpected Loads. *J. Orthop. Sports Phys. Ther.* **2001**, *31*, 708–729. [CrossRef] [PubMed]
14. Hernández-Davó, J.L.; Sabido, R.; Moya-Ramón, M.; Blazevich, A.J. Load Knowledge Reduces Rapid Force Production and Muscle Activation during Maximal-Effort Concentric Lifts. *Eur. J. Appl. Physiol.* **2015**, *115*, 2571–2581. [CrossRef]
15. Hojka, V.; Šťastný, P.; Tufano, J.J.; Omcirk, D.; Janikov, M.T.; Komarc, M.; Jebavý, R. Does a Linear Position Transducer Placed on a Stick and Belt Provide Sufficient Validity and Reliability of Countermovement Jump Performance Outcomes? *Biol. Sport* **2021**, *39*, 341–348. [CrossRef] [PubMed]
16. Cohen, J. *Statistical Power Analysis for the Behavioral Sciences*, 2nd ed.; Routledge: New York, NY, USA, 1988; ISBN 978-0-203-77158-7.
17. Koo, T.K.; Li, M.Y. A Guideline of Selecting and Reporting Intraclass Correlation Coefficients for Reliability Research. *J. Chiropr. Med.* **2016**, *15*, 155–163. [CrossRef]
18. Jidovtseff, B.; Quievre, J.; Nigel, H.; Cronin, J. Influence of Jumping Strategy on Kinetic and Kinematic Variables. *J. Sports Med. Phys. Fit.* **2014**, *54*, 129–138.
19. Pérez-Castilla, A.; Rojas, F.J.; Gómez-Martínez, F.; García-Ramos, A. Vertical Jump Performance Is Affected by the Velocity and Depth of the Countermovement. *Sports Biomech.* **2021**, *20*, 1015–1030. [CrossRef]
20. Pérez-Castilla, A.; Weakley, J.; García-Pinillos, F.; Rojas, F.J.; García-Ramos, A. Influence of Countermovement Depth on the Countermovement Jump-Derived Reactive Strength Index Modified. *Eur. J. Sport Sci.* **2021**, *21*, 1606–1616. [CrossRef]
21. Moran, K.A.; Wallace, E.S. Eccentric Loading and Range of Knee Joint Motion Effects on Performance Enhancement in Vertical Jumping. *Hum. Mov. Sci.* **2007**, *26*, 824–840. [CrossRef]
22. Sánchez-Sixto, A.; Harrison, A.J.; Floría, P. Larger Countermovement Increases the Jump Height of Countermovement Jump. *Sports* **2018**, *6*, 131. [CrossRef]
23. Guess, T.M.; Gray, A.D.; Willis, B.W.; Guess, M.M.; Sherman, S.L.; Chapman, D.W.; Mann, J.B. Force-Time Waveform Shape Reveals Countermovement Jump Strategies of Collegiate Athletes. *Sports* **2020**, *8*, 159. [CrossRef] [PubMed]
24. Barker, L.A.; Harry, J.R.; Mercer, J.A. Relationships between Countermovement Jump Ground Reaction Forces and Jump Height, Reactive Strength Index, and Jump Time. *J. Strength Cond. Res.* **2018**, *32*, 248–254. [CrossRef] [PubMed]
25. Iida, Y.; Kanehisa, H.; Inaba, Y.; Nakazawa, K. Short-Term Landing Training Attenuates Landing Impact and Improves Jump Height in Landing-to-Jump Movement. *J. Strength Cond. Res.* **2013**, *27*, 1560–1567. [CrossRef] [PubMed]
26. Lack, S.; Neal, B.; De Oliveira Silva, D.; Barton, C. How to Manage Patellofemoral Pain—Understanding the Multifactorial Nature and Treatment Options. *Phys. Ther. Sport* **2018**, *32*, 155–166. [CrossRef]
27. Sisk, D.; Fredericson, M. Update of Risk Factors, Diagnosis, and Management of Patellofemoral Pain. *Curr. Rev. Musculoskelet. Med.* **2019**, *12*, 534–541. [CrossRef]

28. Nunes, G.S.; Barton, C.J.; Viadanna Serrão, F. Females with Patellofemoral Pain Have Impaired Impact Absorption during a Single-Legged Drop Vertical Jump. *Gait Posture* **2019**, *68*, 346–351. [CrossRef]
29. Atkins, L.T.; James, C.R.; Yang, H.S.; Sizer, P.S.; Brismée, J.-M.; Sawyer, S.F.; Powers, C.M. Changes in Patellofemoral Pain Resulting from Repetitive Impact Landings Are Associated with the Magnitude and Rate of Patellofemoral Joint Loading. *Clin. Biomech.* **2018**, *53*, 31–36. [CrossRef]

Journal of
Functional Morphology and Kinesiology

MDPI

Article

Differences between Elite and Professional Male Handball Players in Kinematic Parameters of Single Fake Movement

Ante Burger [1], Dario Vrdoljak [2], Nikola Foretić [2,*], Miodrag Spasić [2] and Vladimir Pavlinović [2]

1 University Department of Health Studies, University of Split, 21000 Split, Croatia
2 Faculty of Kinesiology, University of Split, 21000 Split, Croatia; darvrd@kifst.hr (D.V.)
* Correspondence: nikola.foretic@kifst.hr

Abstract: Feint movement is an important factor for offensive players to outplay their guard, and score. So far, there is no evidence of feint biomechanical analysis on a sample of elite players in handball or other team sports. Therefore, this study aimed to investigate kinematic parameters of single side fake movement between elite and professional level handball players. The sample of participants consisted of 10 handball players divided into two subsamples: elite handball players (100.00 ± 8.00 kg; 196.00 ± 4.64 cm) and professional handball players (91.20 ± 3.42 kg; 192.4 ± 7.30 cm). The kinematic analysis was conducted using a GAIT—LaBACS software system. Variables consisted of two phases (fake phase and actual phase) of feint single change of direction. Both phases included seven kinematic parameters that were observed. Statistical analysis included descriptive statistic parameters. The differences between elite and professional handball players were analyzed by multivariate and univariate variance analysis. Results showed significant differences between elite and professional players ($\lambda = 0.44$, $p = 0.00$), in fake phase (i.e., 1. Phase). The results also indicate that in there is no statistically significant difference between both groups ($\lambda = 0.64$, $p = 0.22$). Two variables had significant differences between elite and professional players (i.e., step length of the stride leg ($p = 0.02$) and moving the leg opposite the throwing arm in space ($p = 0.00$)). To conclude, the article examines specific movement patterns of single side fake movement in elite players and the confirmed importance of efficient skill execution in top level handball. On the contrary, less skilled players use more space for the same technical element.

Keywords: team handball; feint movement; biomechanics

Citation: Burger, A.; Vrdoljak, D.; Foretić, N.; Spasić, M.; Pavlinović, V. Differences between Elite and Professional Male Handball Players in Kinematic Parameters of Single Fake Movement. *J. Funct. Morphol. Kinesiol.* **2023**, *8*, 47. https://doi.org/10.3390/jfmk8020047

Academic Editor: Diego Minciacchi

Received: 22 March 2023
Revised: 14 April 2023
Accepted: 18 April 2023
Published: 20 April 2023

1. Introduction

Handball is a technical-tactically, physically and psychologically complex sport that requires full engagement from players [1,2]. Factors that influence the performance of handball players are numerous, because of its complexity and structure. The most important of the factors is an efficient execution of specific tasks, which is an aggregation of running, jumping and direction changes with sport-specific passing, catching, throwing and physical contact [3,4]. Additionally, handball coaches assess a player's offensive quality by his ability to outplay his guard. This is carried out with shoots or fake movements/feints. While shoots are used to finish the attack, feints are the most important elements for gaining an advantage over a defender and making cracks in defense [5]. In modern handball, defensive zone formations are set very deep with strong and explosive defenders [6]. Shooting on goal has become very difficult to achieve just through the individual actions of the attacker. Hence, shooting must precede an advantage that another player gained. Gained advantage may end with a direct shoot of player who performs a feint. Mostly, due to the defensive corrective actions (such as overtaking and helping), an attacker passes the ball and gains advantage during the fake movement, thus transferring the ball to a player with a better position for shooting into the goal. From this reason, it is expected that during the handball match, there are significantly more fake movements than shots on goal.

There are many fake movements that attackers use to trick defenders, but most complex are feints with the ball. These movements are commonly used by handball players, but are lacking any scientific approach. The feint single change of direction belongs to the attacking technique with the ball. Rogulj et al. [7] define feints as a technical element by which a player in attack achieves a temporal or spatial advantage over a defensive player who guards him. This motion allows him to carry out undisturbed realization or ensure continued technical and tactical action in more favorable conditions than the opponent [8]. Feints should be carried out when the defensive players least expect it. Additionally, according to Vuleta et al. [9], center back players' high level of performance is defined by good feint execution.

A review of the aforementioned aspects leads to the assumption that the execution technique of those skills plays an important role in success determination. Moreover, apart from anthropology, motor and functional skills, individual technique plays an important role in differentiation of elite and non-elite players [10–12]. Lanka et al. [13] describe the technique as executive motor-neuromuscular activity and motor realization of imaginary movement in accordance with basic biomechanical principles. It is important to note that the ideal technique may not be the best for every player as they differ from each other in motor and morphological features. Furthermore, some studies showed that difference between non-elite and elite handball players can be found in the use of techniques, especially in throwing motions [14].

So far, the feint technique has been analyzed only as a sport-specific test [15] or in teaching programs for beginners [16]. The importance of feint execution in other team sports was detected in several previous studies, one of which is basketball, where players use one-on-one feints with dribbles and body movements in offensive situations against opponents [17]. Apart from that, dribbling with feints is also used by basketball players [18]. Both studies gave an informational model for performance enhancement in game situations. Moreover, Güldenpenning et al. [19] showed how elite and novice athletes in volleyball differ in feint recognition. Authors concluded that elite athletes were able to predict feint attack at an earlier stage, and therefore showed the importance of good feint execution at high level of sport.

However, there is no evidence of feint biomechanical analysis on elite players sample in handball or other team sports. Following all aforementioned, this study aimed to investigate kinematic parameters of single side fake movement between elite and professional level handball players. More specifically, step lengths, angles between joints and movement speed will be measured by kinematic analysis.

2. Materials and Methods

2.1. Participants

The sample of participants consisted of 10 handball players divided into two subsamples: elite (representative) and quality (first league) players. The selection criteria for elite players was participation in at least five major top-level handball competitions (European Championship, World Championship or Olympic Games). The selection criteria for quality players was participation in at least three seasons in the First Division National Championship. For the sake of time efficiency, all players were recruited from wider surroundings of the city of Split during the offseason period. Elite handball players (representative sample) were represented by five players who played for the Croatian national handball team and won medals in the Olympic Games, World and European handball championships (body mass: 100.00 ± 8.00 kg, body height: 196.00 ± 4.64 cm, four back players and one wing player). Professional handball players (first league sample) were represented by five players who have a minimum of three seasons in their handball career in the First Croatian Handball League (body mass: 91.20 ± 3.42 kg, body height: 192.4 ± 7.30 cm, four back players and one wing player). Informed consent was obtained from all subjects involved in the study. Experimental procedures were completed following the declaration of Helsinki. All athletes participated in handball training on a daily basis, which had a significantly higher risk than the testing procedure conducted. All of them were aware of minimal risk identified and voluntarily participated in the study. Hence, there was no need for ethical board approval.

2.2. Design and Procedures

The kinematic analysis was conducted using a GAIT—LaBACS software system (ver. 1.0, Split, Croatia) designed at the Faculty of Electrical and Mechanical Engineering and Naval Architecture in Split [20]. Variables consisted of two groups that correspond to two phases of feint single change of direction. This element is divided into two parts, the fake phase and the actual (executive) phase. The fake phase had seven kinematic parameters: step length of the stride leg (SLS), the speed of the false part of the phase (TFP), angle of the trunk in relation to the ground (ATG), scrolling the center of gravity of the body-(SCG), duration of the false part of the phase (DFP), moving the leg opposite the throwing arm in space (MOT) and the position of the foot that is opposite of the throwing arm at the end of the actual phase (FOA). The actual (executive) phase had seven kinematic parameters: total length of all steps (TLS), length of step 1 (LS1), length of step 2 (LS2), speed of the first step (S1S), speed of the second step (S2S), direction of step 2 (DS2) and duration of the actual (executive) phase (DAP).

For the purposes of kinematic analysis, twelve symmetric reference points were defined: The reference point is placed at the lower end of the outer ankle (malleoli); the reference point is placed in the middle of the knee joint from the front at the chip position (patella); the reference point is placed at the top of the pelvic bone at the point of the (cresta iliaca); the reference point is set on the (acromion); the reference point is placed at the end of the ulnar bone (olecranon); the reference point is placed at the ends of the ulnar and radial bones from the anterior side (Figure 1). The recording was performed in the sport hall, with a previous calibration of the measuring system, while the layout of the cameras was such that all twelve reference points on the body of the subjects were visible. The cameras used were BASLER-402-FC (Dortmund, Germany), reproducing 100 pictures per second, and PANASONIC VW-D5100 reproducing 50 pictures per second. First of all, the measuring space had to calibrated. Calibration was performed by placing an object of known dimensions in the measurement space. After that, the XY values of the reference points on that object were read (from the camera images). By entering the read values from the images, as well as the spatial coordinates of reference points, it is possible to use the DLT (direct linear transformation) method to form a matrix used to calculate spatial coordinates for every other point in the measurement space. 3D calibration of the space was performed by placing eight static markers, where four were positioned in the vertical direction on a metal rod with a defined height, and four positioned on the ground, arranged at the bottom of the metal rod in a proportional square spacing. Video sequences were reviewed, and reference point positions were read and recorded from each camera. Using the DLT method, the spatial position of the reference points at each time point of the element design was calculated. For each analyzed element, the recording was repeated three times. The analysis of the quality of the recordings was carried out immediately after the testing, so that any errors in the video could be corrected or repeated with a new test.

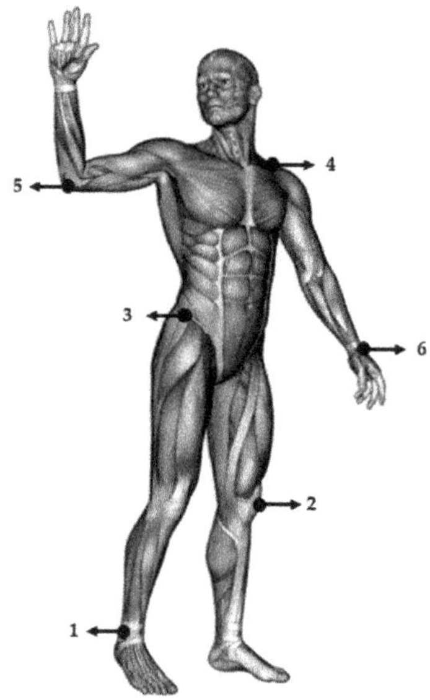

Figure 1. Reference points for kinematic analysis, defined as: (1) malleoli; (2) patella; (3) crista iliaca; (4) acromion; (5) olecranon; (6) end of ulnar and radial bones.

2.3. Statistical Analysis

Statistical analysis included calculation of descriptive statistic parameters while normality was tested using the Kolmogorov–Smirnov test procedure. The differences in kinematic parameters between elite and professional handball players were analyzed by multivariate (used to determine differences between all parameters of the given fake movement phase) and univariate variance analysis (used to determine differences between each parameter of the given fake movement phase). The level of significance was set at $p < 0.05$. The software Statistica ver.13.0 (Dell Inc., Austin, TX, USA) was used for all analysis.

3. Results

The Kolmogorov–Smirnov test exhibited normal distribution of the variables ($p > 0.20$). The calculation of descriptive statistic parameters included means, standard deviations and minimum and maximum for both phases and groups.

Table 1 shows statistically significant differences between elite and professional handball players ($\lambda = 0.44$, $p = 0.00$), in fake phase (i.e., 1. Phase). The results also indicate that in the second statistically significant difference between groups, do not exist ($\lambda = 0.64$, $p = 0.22$).

Table 1. Differences between elite and professional handball players in parameters of single fake movement calculated by MANOVA.

1. Phase–fake phase			
Test	Wilks λ	F	*p*
λ	0.44	4.64	0.00 *
2. Phase–actual (executive phase)			
Test	Wilks λ	F	*p*
λ	0.64	1.5	0.22

λ—Wilk's lambda, F—F test, *p*—Coefficient of significant difference, *—$p < 0.05$.

Univariate analysis of variance (ANOVA) is presented in Table 2. From the given results it is visible that two variables had significant difference between elite and professional players (i.e., SLS (p = 0.02) and MOT (p = 0.00)). Elite players execute smaller steps (SLS) during the first phase of the fake movement (106.44 cm) than professional players (133.55 cm). MOT results implicate that elite players exhibit a lesser angle (17.70°) than professional players (31.42°).

Table 2. Differences between elite and professional handball players in the first phase of fake movement calculated by ANOVA.

	Mean Elite Players	Mean Professional Players	F	p
SLS	106.44 (cm)	133.55 (cm)	5.81	0.02 *
TFP	5.16 (m/s)	5.76 (m/s)	1.13	0.30
ATG	74.35 (°)	75.41 (°)	0.18	0.67
SCG	78.12 (cm)	93.17 (cm)	3.6	0.07
DFP	0.21 (ms)	0.25 (ms)	2.17	0.15
MOT	17.70 (°)	31.42 (°)	11.87	0.00 *
FOA	95.00 (°)	89.00 (°)	3.33	0.07

SLS—step length of the stride leg, TFP—the speed of the false part of the phase, ATG—angle of the trunk in relation to the ground, SCG—scrolling the center of gravity of the body, DFP—duration of the false part of the phase, MOT—moving the leg opposite the throwing arm in space, FOA—the position of the foot that is opposite of the throwing arm at the end of the fake phase, F—F test, p—Coefficient of significant difference, *—$p < 0.05$.

Differences between groups in the second phase of the fake movement are presented in Table 3. In the variables of the second phase, there are no statistically significant differences recorded. However, some differences can be noticed. Specifically, elite players have more speed (4.44 m/s) in this part of the second phase than professionals (4.11 m/s). Additionally, in DS2, elite players have smaller angle of step (13.43°) than professional players (17.64°).

Table 3. Differences between elite and professional handball players in the second phase of the fake movement calculated by ANOVA.

Variables	Mean Elite Players	Mean Professional Players	F	p
TLS	328.44 (cm)	342.78 (cm)	0.54	0.46
LS1	150.95 (cm)	162.08 (cm)	0.76	0.38
LS2	177.56 (cm)	180.51 (cm)	0	0.95
S1S	5.41 (m/s)	5.41 (m/s)	0.01	0.91
S2S	4.44 (m/s)	4.11 (m/s)	1.96	0.17
DS2	13.43 (°)	17.64 (°)	2.48	0.13
DAP	0.77 (ms)	0.80 (ms)	0.24	0.62

TLS—total length of all steps, LS1—length of step 1, LS2—length of step 2, S1S—speed of the first step, S2S—speed of the second step, DS2—direction of step 2, DAP—duration of the actual (executive) phase, F—F test, p—Coefficient of significant difference.

4. Discussion

This study aimed to investigate kinematic parameters of single side fake movement between elite and professional handball players. The obtained results indicate several important findings: (1) Players differ in the first phase of the feint; (2) elite players have a smaller first step and move more sideways in the first phase; and (3) differences in the second phase are connected to first phase performance but are not statistically significant.

The literature review refers to significant differences in movement pattern efficiency dependable on players' skill levels. This was noticed between skilled and novice players in volleyball, soccer, basketball and tennis [17,21–25]. For example, Fujii, Yamada and Oda [21] demonstrated that skilled basketball players had smaller decrement in maximal sprint to maximal dribbling speed. Their results indicate greater compensation in dribbling through body segment movements. Moreover, basketball players also showed differences in ball handling [22]. It was shown that skilled players have longer and more consistent ball contact and, therefore, better spatial control of the ball. According to Loffing and Hagemann (2014), unlike the novice, skilled tennis players consider the reliability of different information sources by weighting the available contextual and kinematic cues differently in the course of an opponent's unfolding action [25]. Furthermore, studies conducted in soccer show effective upper-body movement in skilled players to be a key factor in creating better initial conditions for a more explosive muscle contraction during kicking [24]. Obviously, it could be expected that, in sports, a higher level of sport competition demands more efficient technique and movement pattern performance. Efficient movement saves time, increases speed and power and therefore represents advantage to more efficient athletes [26]. Movement pattern efficiency and its dependence on skill level in handball was investigated in several studies. Those studies analyzed kinematic parameters and were mainly focused on overarm throwing and jump-shot performance [27–30]. In some of these studies, a lack of differences was observed between novice and expert players [27]. The authors showed that changing the goal of the task similarly effected velocity of the ball and movement of body segments in both groups. Therefore, authors concluded that, in this particular case, training experience is not related to overarm throwing performance. However, in jump-shots, technique differences were found. Elite players perform jump-shots differently. They had increased trunk flexion and rotational angular velocity, which results in an increase in ball release speed [28]. Making a large step (like professional players in this study) lower than his center of mass causes a wider angle of equilibrium. A wider angle of equilibrium implies better stability, which can be counterproductive in this situation after which an explosive countermovement is required. Furthermore, by taking such a large step, one put his lower extremities in a disadvantageous position for the next quick reaction, considering the angles in the joints and torques that the muscles can produce at these angles. Not to mention that a shorter step also takes less time to perform, and it is known that time is a significant factor when performing a successful feint.

Our results correspond to previously mentioned studies in terms of differences between elite and professional players, as some of kinematic parameters differ between the groups. Still, it was evident only in the first phase of the fake movement. Results show that elite players performed specific movement pattern with the shorter sidestep for 27.11 cm than professional players.

Additionally, elite players conduct fake movements more to the side, unlike professional players who perform them more diagonally. This can be noticed from the MOT variable in which professional players during the fake movement have a larger angle of 13.72° between the feet and frontal plane. The diagonal position of the attackers' feet puts the offensive player in less favorable position towards the defender. This way the attacker is closer to defender and has less space for executive phase of fake movement that should be performed faster and more explosively. In addition, by thrusting too far forward, the attacker leaves the defender more sideways, and in order to pass him on the other side in the next moment (which is the next phase of the feint), it is necessary to go "back" and then go around the defender. Again, we see that time is important for performing a successful feint. The defender's job is to stop the attacker with tackles. In those tackles, he always tries to be as close as possible to the attacker [31]. Hence, we can conclude that elite players' movements in the first phase is more efficient in real game situations. Moving the leg more sideways allows the player to shift his center of gravity without losing balance while also maintaining the strong support that is necessary for the agile and explosive second phase of feint [32]. Shifting the center of gravity is important for the reaction of the defender because an experienced defender should react only to a significant change in the position of the center of gravity of attacker and not to feinting (only) with the extremities. This is supported by the fact that the best representation of the body is its center of gravity If one wants to describe trajectory of body's movement in general, he or she can use that specific point which would substitute the whole body [33] A bigger shift sideways forces the defender to react with a more resolute tackle. Tackles of this kind are more advantageous for defenders, since they have to move more and faster. Rapid dropping out of defending position is critical for defensive stance, body balance and consequently losing the duel with the attacker. Although statistical analysis did not show significant differences between the two groups, it is evident that elite players use less space in the second phase of the feint. Specifically, total length of all steps, length of the first and second step separately, are shorter in elite than in professional players, respectively. Efficiency of the second phase is directly influenced by first phase and, most probably, by first phase execution. As stated earlier, the elite players perform it in a smaller space than professional players. Hence, they need less space and distance (and consequently time) to finish it in the executive phase of the feint. Overall, it can be stated that elite players perform single side fake movement more efficiently than professional players.

This study had several considerable limitations. Although players in two groups played the same playing position, we noticed a significant difference in anthropometrical indices that could influence skill execution. In several studies, positive correlations were noticed between the body mass and body height and throwing kinematic parameters. Body mass influenced speed variables while body height influenced the height of the body's center of gravity [10,28]. Similar influences are expected in our sample and contribute to the possible distribution of data homogeneity.

Additionally, lack of some more precise anthropometric measurements could be influential in the assessment of feint movement. Furthermore, in this study we have researched only kinematic parameters of single side fake movement. Future studies should include other factors that are important in handball players' feinting performance. Two aspects should be specially researched: (1) cognitive abilities of players, such as perceptual and decision-making factors, and (2) physical abilities of players, such as timing, muscle quality, strength, power and/or reactive agility factors. It would be important to know which of these to is more important for successful handball feint performance.

5. Conclusions

Research of kinematic parameters of single side fake movement for feint single change of direction has not been carried out so far, not just in handball but in team sport games generally. Hence, it was impossible to compare the obtained results and methodology with literature review. Results show specific movement pattern of single side fake movement in elite players and confirmed importance of efficient skill execution in top level handball. On the contrary, less skilled players use more space for the same technical element. Results of the study could direct handball coaches in better understanding single side fake movement performance. During teaching this important element they should focus on: (1) optimal step length; players' needs to adapt first step length to a defender's position and his anthropometry, (2) maintaining a straight body position while shifting center of gravity; during teaching the single side fake movement strait body position is crucial for controlling body balance and good perception of the game and (3) controlling the distance from defender; conducting single side fake movement to close to defender is inefficient, so the attacker should be learned to avoid physical contact and perform it on 5 to 10 cm distance of the defender's arms reach.

Author Contributions: Conceptualization, A.B. and N.F.; methodology, D.V.; software, A.B.; validation, A.B., D.V. and N.F.; formal analysis, M.S.; investigation, A.B. and N.F.; resources, A.B.; data curation, V.P.; writing—original draft preparation, D.V., N.F. and M.S.; writing—review and editing, A.B. and V.P.; visualization, V.P.; supervision, A.B.; project administration, M.S.; funding acquisition, A.B. All authors have read and agreed to the published version of the manuscript.

Funding: This research received no external funding.

Institutional Review Board Statement: All athletes, on daily basis participated in handball training, which has significantly higher risk than testing procedure conducted. All of them were aware of minimal risk identified and voluntary participated in the study. Hence, there was no need for ethical board approval.

Informed Consent Statement: Informed consent was obtained from all participants involved in the study.

Conflicts of Interest: The authors declare no conflict of interest.

References

1. Rogulj, N. Differences in situation-related indicators of the handball game in relation to the achieved competitive results of teams at 1999 World Championship in Egypt. *Kinesiology* **2000**, *32*, 63–74.
2. Wagner, H.; Finkenzeller, T.; Würth, S.; von Duvillard, S.P. Individual and Team Performance in Team-Handball: A Review. *J. Sports Sci. Med.* **2014**, *13*, 808–816. [PubMed]
3. Michalsik, L.; Aagaard, P.; Madsen, K.J. Locomotion characteristics and match-induced impairments in physical performance in male elite team handball players. *Int. J. Sports Med.* **2013**, *34*, 590–599. [CrossRef] [PubMed]
4. Póvoas, S.C.A.; Seabra, A.F.T.; Ascensão, A.A.M.R.; Magalhães, J.; Soares, J.M.C.; Rebelo, A.N.C. Physical and physiological demands of elite team handball. *J. Strength Cond. Res.* **2012**, *26*, 3365–3375. [CrossRef] [PubMed]
5. Fasold, F.; Meyer, J.; Klatt, S. Effects of throwing feints on attack strategy in male elite handball: A post-hoc video analysis. *Int. J. Perform. Anal. Sport* **2023**, *1*, 1–12. [CrossRef]
6. Karcher, C.; Buchheit, M. On-court demands of elite handball, with special reference to playing positions. *Sports Med.* **2014**, *44*, 797–814. [CrossRef]
7. Rogulj, N.; Papić, V.; Zanchi, V. Leg abduction kinematic characteristics of handball goalkeeper. In Proceedings of the 6th Annual Congress of the European College of Sport Science, Cologne, Germany, 24–28 July 2001.
8. Burger, S.; Cataliotti, F.; Fort, C.; Minardi, F.; Inguscio, M.; Chiofalo, M.L.; Tosi, M. Burger et al. Reply: Dynamical or Landau instability? *Phys. Rev. Lett.* **2002**, *89*, 088902. [CrossRef]
9. Vuleta, D.; Milanović, D.; Rataj, I. Vrednovanje tehničko-taktičkih elemenata srednjih vanjskih napadača u rukometu. In Proceedings of the 16. Ljetna Škola Kineziologa Republike Hrvatske, Poreč, Croatia, 19–23 July 2007; pp. 554–560.
10. Ohnjec, K.; Vuleta, D.; Milanović, D.; Gruić, I. Performance indicators of teams at the 2003 World Handball Championship for Women in Croatia. *Kinesiology* **2008**, *40*, 69–79.
11. Trninić, S.; Dizdar, D. System of the performance evaluation criteria weighted per positions in the basketball game. *Coll. Antropol.* **2000**, *24*, 217–234.
12. Massuça, L.M.; Fragoso, I.; Teles, J. Attributes of Top Elite Team-Handball Players. *J. Strength Cond. Res.* **2014**, *28*, 178–186. [CrossRef]
13. Lanka, J.; Zatsiorsky, V. Biomechanics in Sport. In *Performance Enhacement and Injury Prevention*; Blackwell Science: Oxford, UK, 2000; pp. 435–457.
14. Moss, S.L.; McWhannell, N.; Michalsik, L.B.; Twist, C. Anthropometric and physical performance characteristics of top-elite, elite and non-elite youth female team handball players. *J. Sports Sci.* **2015**, *33*, 1780–1789. [CrossRef] [PubMed]
15. Fasold, F.; Houseman, L.; Noel, B.; Klatt, S.J.H.M. Handball-specific skill acquisition by use of different instruction methods. *Hum. Mov.* **2020**, *22*, 45–53. [CrossRef]
16. Henigman, D.; Ohnjec, K.J.A.K. Teaching feints to handball beginners: Preliminary results. *Ann. Kinesiol.* **2021**, *12*, 5–13. [CrossRef]
17. Hashizume, J.; Ae, M.; Otsu, T. A biomechanical analysis of 1-on-1 offense's feint motion with a dribble in basketball. In Proceedings of the ISBS-Conference Proceedings Archive, Tsukuba, Japan, 18–22 July 2016.
18. Hu, B.; Dong, Z. Biomechanical analysis on turnaround dribble in basketball. *Rev. Ibérica Sist. E Tecnol. Inf.* **2016**, *17b*, 242–254.
19. Güldenpenning, I.; Steinke, A.; Koester, D.; Schack, T. Athletes and novices are differently capable to recognize feint and non-feint actions. *Exp. Brain Res.* **2013**, *230*, 333–343. [CrossRef]
20. Blažević, Z.; Cecić, M.; Grujić-Šupuk, T.; Papić, V.; Pavelin, A.; Zanchi, I.; Zanchi, V. Dislocated Measurements of Human Gait and Central Data Processing. In Proceedings of the SoftCOM 2001, Split, Croatia, 10–14 October 2001.
21. Fujii, K.; Yamada, Y.; Oda, S. Skilled Basketball Players Rotate Their Shoulders More during Running While Dribbling. *Percept. Mot. Ski.* **2010**, *110*, 983–994. [CrossRef]
22. Katsuhara, Y.; Fujii, S.; Kametani, R.; Oda, S. Spatiotemporal Characteristics of Rhythmic, Stationary Basketball Bouncing in Skilled and Unskilled Players. *Percept. Mot. Ski.* **2010**, *110*, 469–478. [CrossRef]
23. Garcia, S.; Delattre, N.; Berton, E.; Divrechy, G.; Rao, G. Comparison of landing kinematics and kinetics between experienced and novice volleyball players during block and spike jumps. *BMC Sports Sci. Med. Rehabil.* **2022**, *14*, 105. [CrossRef]
24. Shan, G.; Westerhoff, P. Soccer: Full-body kinematic characteristics of the maximal instep soccer kick by male soccer players and parameters related to kick quality. *Sports Biomech.* **2005**, *4*, 59–72. [CrossRef]
25. Loffing, F.; Hagemann, N. On-court position influences skilled tennis players' anticipation of shot outcome. *J. Sport Exerc. Psychol.* **2014**, *36*, 14–26. [CrossRef]
26. Brewer, C. *Athletic Movement Skills: Training for Sports Performance*; Human Kinetics: North Yorkshire, UK, 2017.
27. Van den Tillaar, R.; Ettema, G. A Comparison between Novices and Experts of the Velocity-Accuracy Trade-Off in Overarm Throwing. *Percept. Mot. Ski.* **2006**, *103*, 503–514. [CrossRef] [PubMed]

28. Wagner, H.; Buchecker, M.; von Duvillard, S.P.; Müller, E. Kinematic description of elite vs. Low level players in team-handball jump throw. *J. Sports Sci. Med.* **2010**, *9*, 15–23. [PubMed]
29. Schorer, J.; Baker, J.; Fath, F.; Jaitner, T. Identification of interindividual and intraindividual movement patterns in handball players of varying expertise levels. *J. Mot. Behav.* **2007**, *39*, 409–421. [CrossRef] [PubMed]
30. Wagner, H.; Pfusterschmied, J.; Klous, M.; von Duvillard, S.P.; Müller, E. Movement variability and skill level of various throwing techniques. *Hum. Mov. Sci.* **2012**, *31*, 78–90. [CrossRef] [PubMed]
31. Michalsik, L.B.; Madsen, K.; Aagaard, P. Technical match characteristics and influence of body anthropometry on playing performance in male elite team handball. *J. Strength Cond. Res.* **2015**, *29*, 416–428. [CrossRef] [PubMed]
32. Bradshaw, R.J.; Young, W.B.; Russell, A.; Burge, P. Comparison of offensive agility techniques in Australian Rules football. *J. Sci. Med. Sport* **2011**, *14*, 65–69. [CrossRef] [PubMed]
33. Erdman, W.S. Center of mass of the human body helps in analysis of balance and movement. *MOJ Appl. Bionics Biomech.* **2018**, *2*, 144–148.

Journal of
Functional Morphology and Kinesiology

MDPI

Case Report

Use of Robot-Assisted Ankle Training in a Patient with an Incomplete Spinal Cord Injury: A Case Report

Kazunori Koseki [1], Kazushi Takahashi [1], Satoshi Yamamoto [2], Kenichi Yoshikawa [1], Atsushi Abe [1] and Hirotaka Mutsuzaki [3,*]

1 Department of Physical Therapy, Ibaraki Prefectural University of Health Sciences Hospital, 4733 Ami, Inashiki-gun, Ibaraki 300-0331, Japan
2 Department of Physical Therapy, Ibaraki Prefectural University of Health Sciences, 4669-2 Ami, Inashiki-gun, Ibaraki 300-0394, Japan
3 Department of Orthopaedic Surgery, Ibaraki Prefectural University of Health Sciences Hospital, 4733 Ami, Inashiki-gun, Ibaraki 300-0331, Japan
* Correspondence: mutsuzaki@ipu.ac.jp; Tel.: +81-29-888-4000

Abstract: Rehabilitation interventions are crucial in promoting neuroplasticity after spinal cord injury (SCI). We provided rehabilitation with a single-joint hybrid assistive limb (HAL-SJ) ankle joint unit (HAL-T) in a patient with incomplete SCI. The patient had incomplete paraplegia and SCI (neurological injury height: L1, ASIA Impairment Scale: C, ASIA motor score (R/L) L4:0/0, S1:1/0) following a rupture fracture of the first lumbar vertebra. The HAL-T consisted of a combination of ankle plantar dorsiflexion exercises in the sitting position, knee flexion, and extension exercises in the standing position, and stepping exercises in the standing position with HAL assistance. The plantar dorsiflexion angles of the left and right ankle joints and electromyograms of the tibialis anterior and gastrocnemius muscles were measured and compared using a three-dimensional motion analyzer and surface electromyography before and after HAL-T intervention. Phasic electromyographic activity was developed in the left tibialis anterior muscle during plantar dorsiflexion of the ankle joint after the intervention. No changes were observed in the left and right ankle joint angles. We experienced a case in which intervention using HAL-SJ induced muscle potentials in a patient with a spinal cord injury who was unable to perform voluntary ankle movements due to severe motor–sensory dysfunction.

Keywords: incomplete spinal cord injury; robot-assisted training; robotics

Citation: Koseki, K.; Takahashi, K.; Yamamoto, S.; Yoshikawa, K.; Abe, A.; Mutsuzaki, H. Use of Robot-Assisted Ankle Training in a Patient with an Incomplete Spinal Cord Injury: A Case Report. *J. Funct. Morphol. Kinesiol.* **2023**, *8*, 31. https://doi.org/10.3390/jfmk8010031

Academic Editor: Diego Minciacchi

Received: 7 January 2023
Revised: 24 February 2023
Accepted: 24 February 2023
Published: 27 February 2023

1. Introduction

Traumatic spinal cord injury (SCI) is caused by physical impact on the spinal cord due to several reasons, such as road traffic trauma and falls from heights. In the United States, the estimated incidence rate of SCI is 52–54 cases per million people [1,2], while in Japan, it is approximately 49 cases per million people [3]. The primary pathology of SCI is damage to the spinal cord due to primary trauma and subsequent secondary injury, which results in the blockage of descending and ascending nerve conduction from the brain, resulting in motor and sensory deficits, bladder–rectal deficits, and autonomic neuropathy, following the injury [4]. This causes dysfunction in the ankle joint, which is dominated by the lumbar and sacral spinal cord regions, one of the most vulnerable areas to impairment. Physical and occupational therapies, including mobility exercises using a wheelchair and activities of daily living exercises, are commonly used in patients with paraplegia due to SCI in the lumbar region [5]. In addition, clutches and lower limb orthoses are generally used to compensate for missing ankle joint function during standing and walking practice [5]. In recent years, because of advances in robotics, the effectiveness of interventions using various exoskeletal robots has been verified [6,7]. However, most of these assist with walking movements, and there have been few reports on robotic interventions specifically designed to improve ankle joint function.

The hybrid assistive limb (HAL) has a bioelectrical signal (BES)-based control system that can be assisted by a joint torque driven voluntarily by the wearer. The single-joint-type HAL (HAL-SJ, Cyberdyne, Inc., Tsukuba, Japan) is a robot that can support flexion and extension movements of various joints. Previously, interventions for elbow and knee joint dysfunction have been reported [8–11], but with the expansion of the ankle joint unit, it is being explored for ankle joint dysfunction [12]. Furthermore, the use of HAL-SJ has been reported to improve the function of elbow joint muscles that failed to contract voluntarily in patients with cervical cord injury (C4 level) [13]. Thus, repetition of voluntary exercise using HAL-SJ can improve paralyzed muscles in patients with spinal cord injuries. Hence, this study aimed to perform an intervention (HAL-T) using a single-joint HAL ankle joint unit in a patient with paraplegia due to SCI.

2. Case Report

2.1. Participant

The study patient was a 34-year-old man (height, 169 cm; weight, 79.4 kg). He only had pre-existing medical history of hyperlipidemia. He sustained a burst fracture of the first lumbar vertebra, dislocation of the right shoulder joint due to a fall from a height. At the time of emergency transport, the patient was found to have bilateral lower extremity paralysis and cysto-rectal disturbance, and was diagnosed as having a spinal cord injury. As assessed by The International Standards for Neurological Classification of Spinal Cord Injury, the neurological level of injury at the time of injury was the second lumbar (L1) level, and the American Spinal Cord Injury Association (ASIA) Impairment Scale was B (sensory but not motor function is preserved below the neurological level and includes the sacral segments S4–S5, and no motor function is preserved at more than three levels below the motor level on either side of the body) [14]. The patient was transferred to an acute-care hospital and underwent posterior spinal decompression fusion on the same day (Figure 1a,b). Then, at 18 days after the injury, he was transferred to our hospital for continued rehabilitation. Physical and occupational therapy was initiated at the hospital, focusing on muscle training, standing exercises, walking exercise with orthosis, and activities of daily living (ADL) exercises. On the 101st day post-injury, the patient's neurological injury level was L1, the ASIA Impairment Scale was C, and the ASIA motor score (R/L) was as follows: L2: 4/4, L3: 3/4; L4: 0/0; L5: 0/0; and S1: 1/0; the patient had symptoms of paraplegia. The patient was able to perform basic ADL in the hospital using a wheelchair and was able to walk with a walker using a knee–ankle foot orthosis on both the right and left sides. However, the patient's ankle joint function remained severely impaired, and the patient was highly concerned about recovering the function. Therefore, an intervention using a single-joint HAL ankle unit was performed, in addition to the aforementioned usual rehabilitation programs to improve the ankle joint function.

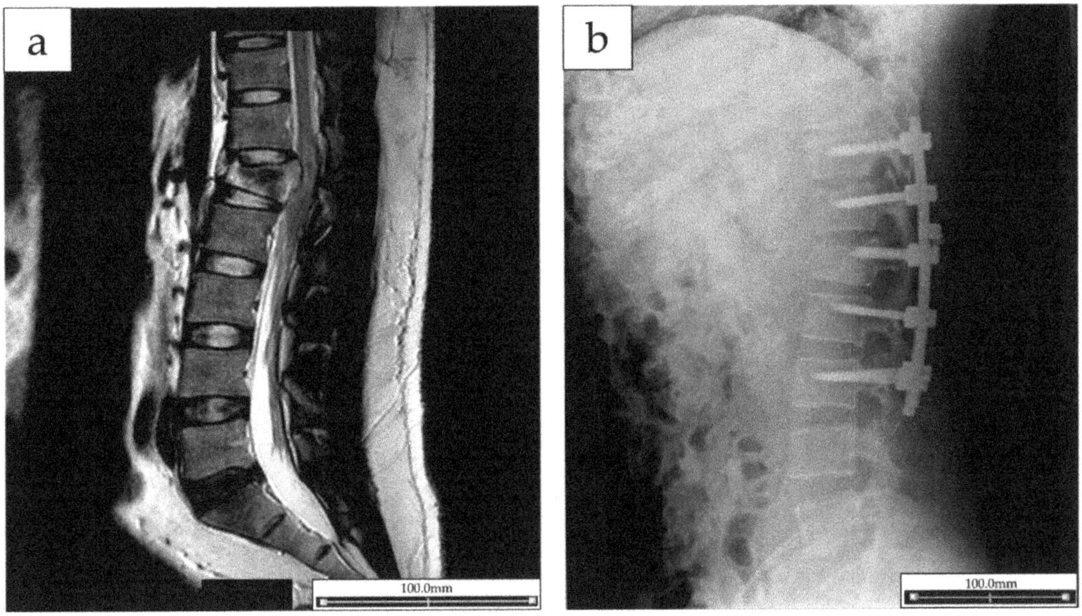

Figure 1. (**a**) Magnetic resonance imaging images of the lumbar spine and spinal cord taken on the date of injury. (**b**) X-ray image after posterior spinal decompression fusion.

2.2. Intervention Used

HAL-SJ allows the wearer to voluntarily perform active assisted exercises through BES-based control. The HAL-SJ electrodes were affixed to the tibialis anterior (TA) and the gastrocnemius (GAS) muscles and driven based on their BES. HAL-SJ was applied to both ankle joints. The HAL-SJ intervention consisted of a combination of (1) ankle joint plantar dorsiflexion exercises in the end-sitting position (Figure 2a), (2) squat exercise, and (3) stepping exercises (Figure 2b) under HAL assistance, 20 min per session, five times a week, 10 times in total. In each intervention, the patient first performed ankle plantar dorsiflexion exercises in the end-sitting position to activate foot muscle activity and movement, followed by squatting and stepping tasks in the standing position. During squat and stepping exercises, a harness was used for fall prevention and partial unloading. Sensitivity adjustment of the amount of assist torque according to the degree of patient BES can be controlled by operating the controller and can be increased or decreased by the control item "assist gain." To counteract undesirable assist due to abnormal BES caused by the antagonist muscles when using HAL-SJ, a control mechanism called "assist balance" can be used to reduce the BES in 20 steps from 0% to 100% in 5% increments. This control item makes it possible to adjust the balance between flexion and extension assist torques. In this intervention, the assist gain and balance were adjusted to provide the desired exercise according to the level of spasticity and BES. The assist gain and balance were changed on a case-by-case basis to allow HAL-SJ to effectively perform joint exercises.

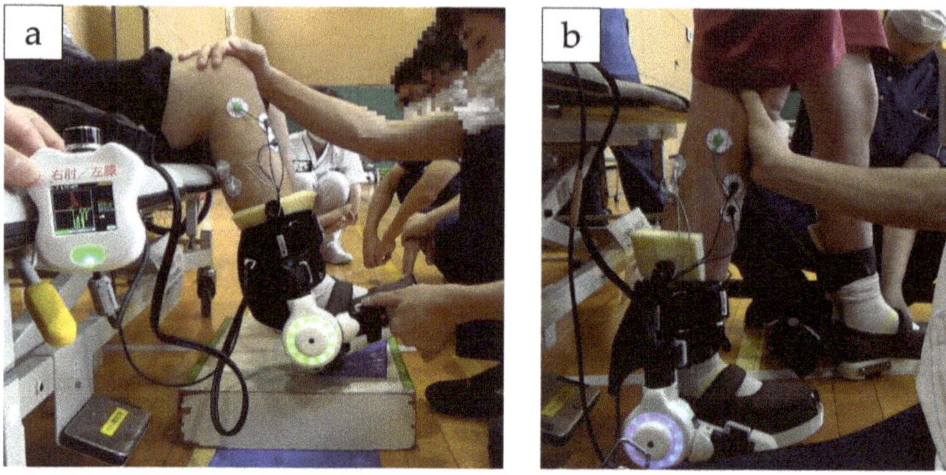

Figure 2. (**a**) Ankle joint plantar dorsiflexion exercises with HAL-SJ in end-sitting position. (**b**) Stepping exercises with HAL-SJ. During squat and stepping exercises, a harness was used to prevent falls and partial unloading.

2.3. Outcome Measurement

Lower limb kinematic parameters were measured, and electromyography (EMG) was performed before and after the HAL-SJ intervention period. The voluntary ankle joint plantar-dorsal flexion movements in the end-sitting position (after 10 s of rest, 10 times each of alternating plantar and dorsiflexion movements at 1 Hz timing, signaled using a metronome) were measured. Both ankle kinematic parameters and EMG were measured using a wireless inertial measurement unit (IMU) system (myoMOTION, Noraxon USA Inc., Scottsdale, AZ, USA) consisting of an IMU, and an Ultium-EMG sensor system (Noraxon Inc.) with a sampling frequency of 2000 Hz, and a bandpass filter of 10–450 Hz. IMUs were placed on the anterior surfaces of the tibia and metatarsals bilaterally to measure ankle joint angles. Each IMU had a local coordinate system and measured acceleration. The joint angles between the IMUs were calculated by the IMU system software (myo RESERCH 3.16.86, Noraxon USA Inc., Scottsdale, AZ, USA). Lower limb kinematic parameters were resampled to 100 points of joint motion from plantar flexion to dorsiflexion and plantar flexion and averaged 10 times. EMG patterns of the bilateral TA and GAS muscles were recorded, and the raw waveform of the EMG was depicted. Then, the EMG data were rectified, smoothed (RMS algorithm with a smoothing window width of 100 ms), added, and averaged for 10 trials.

2.4. Outcome of the Intervention

A total of 10 sessions was performed, and no adverse events, such as abrasions, excessive muscle fatigue, or pain, were observed. In the first session, effective joint movements in the intended direction did not occur sufficiently in the end-sitting lower-limb movements. However, after the second session, joint movements under HAL-assist gradually appeared after providing positive feedback when the intended movements occurred, by visually checking the EMG graph on the HAL controller, adjusting "assist gain" and "assist balance".

Figures 3–5 show the results of lower-limb kinematic parameters and EMG during voluntary ankle joint plantar-dorsal flexion movement in the end-sitting position. Lower limb kinematic parameters showed that there was no obvious phasic joint angle change according to ankle joint plantar dorsiflexion timing on either side before or after the HAL-SJ intervention (Figure 3). Raw EMG data during end-sitting ankle plantar dorsiflexion movements showed no obvious muscle activity in the TA, bilaterally, before the intervention;

however, after the intervention, muscle activity was observed in the left TA (Figure 4). Furthermore, when averaged over 10 plantar dorsiflexion exercises, phasic muscle activity was clearly observed in the left TA during dorsiflexion exercises (Figure 5). No clear muscle activity was observed on either side of the GAS after the intervention.

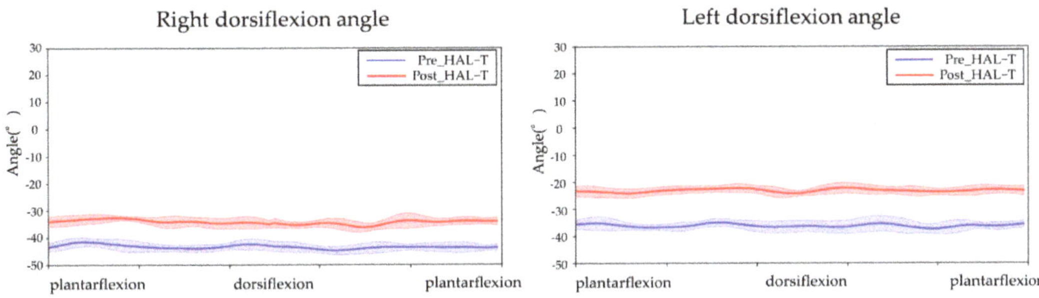

Figure 3. Ankle joint angle during ankle joint plantar dorsiflexion movement in end-sitting position. The blue line shows angle before HAL-T and the red line shows after HAL-T. The 10 ankle plantar dorsiflexion movements were averaged, and the results are shown as means ± standard deviations.

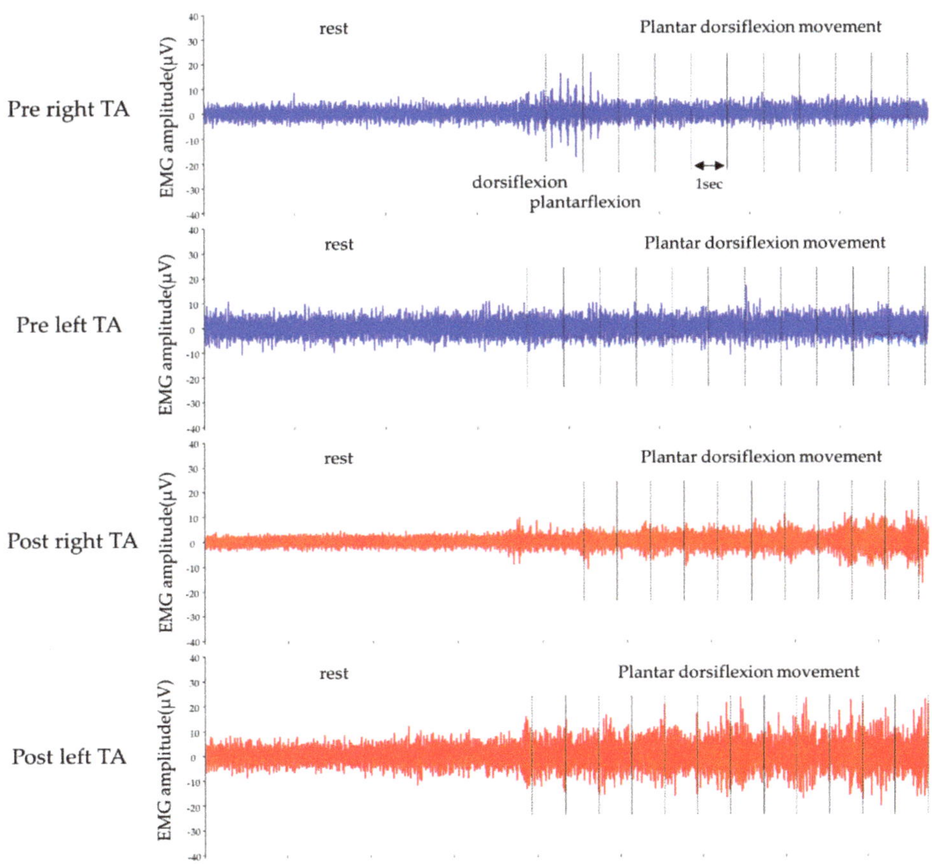

Figure 4. Raw EMGs of the left and right tibialis anterior (TA) muscles during ankle joint plantar-dorsiflexion movement in the end-sitting position.

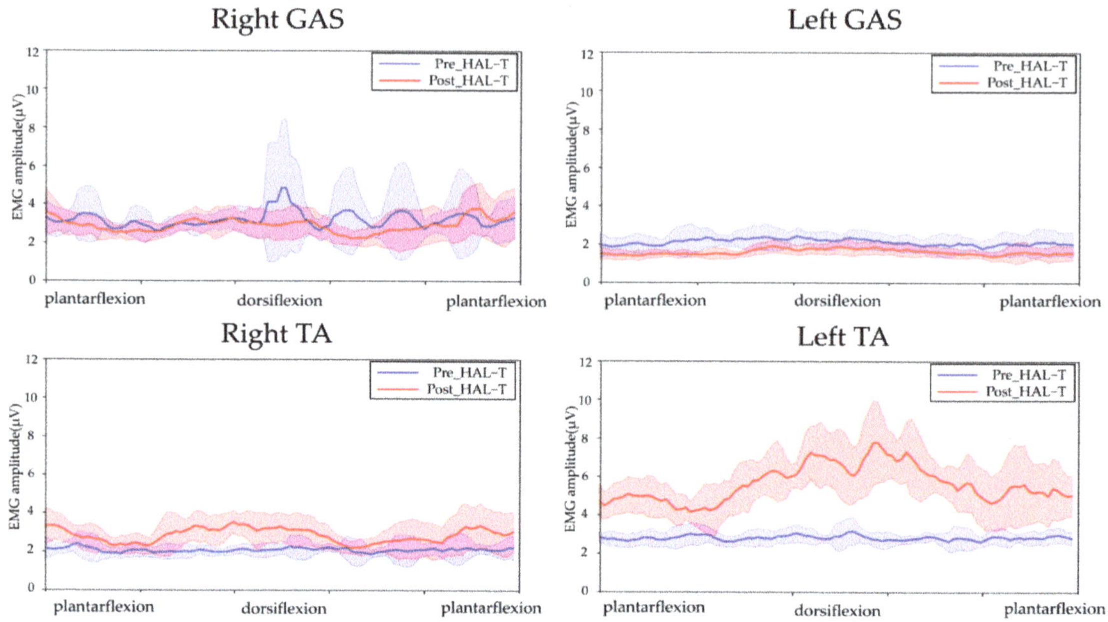

Figure 5. EMGs of the left and right tibialis anterior (TA) and gastrocnemius (GAS) muscles during ankle joint plantar dorsiflexion movement in the end-sitting position. The blue line shows EMG before HAL-T and the red line shows after HAL-T. The raw EMGs were rectified and smoothed (RMS algorithm with a smoothing window width of 100 ms), and the 10 ankle plantar dorsiflexion movements were averaged. The results are presented in the graph as means ± standard deviations.

3. Discussion

The patient had suffered a lumbar fracture and spinal cord injury due to a fall from a height, and had made progress in acquiring ADLs using a wheelchair through rehabilitation. He was just beginning to practice walking with a lower limb orthosis. However, the ankle joint function had not improved sufficiently, and the patient desired further intervention. For this reason, we attempted to provide intervention using the HAL-SJ, an exoskeleton-type device driven by BES. A total of 10 HAL-SJ intervention sessions were performed, and no adverse events, such as abrasions, excessive muscle fatigue, or pain, were observed. Comparison of EMG before and after intervention revealed phasic muscle activity in the left tibialis anterior muscle. However, the current interventions did not produce joint angle changes during ankle joint motion, and did not reach the point where changes in walking ability or ADL occurred.

In this case, no obvious joint movement was observed in the cybernic voluntary control mode driven by the BES at the start of the intervention. However, by adjusting the assist gain, assist level, and assist balance of the HAL to create a state where the joint movement was likely to occur, by visually presenting the EMG graph on the controller and providing positive feedback when the intended joint movement occurred, the voluntary movement gradually appeared. Training duration, high intensity, and augmented feedback are among the factors that have been reported to influence training effectiveness after SCI [15]. In addition to electrical stimulation and biofeedback therapy [16–19], the effectiveness of various types of robotic training [7] has also been reported in patients with SCI and sensorimotor impairment. The concept of plasticity-based functional training emphasizes that more intensive training with adjusted difficulty can be performed using a rehabilitation robot [20,21] and that robotic support should be minimized to challenge the patient's efforts [10,22]. In the present case, the patient had severely impaired ankle sensory–motor function, and without HAL-SJ there was no clear ankle motion. Even though the patient did not sense movements, it was possible to use the equipment to generate foot motion and perform repetitive training at an optimal load with HAL-SJ. The intervention with the HAL-SJ contributed to obtaining visual and sensory feedback, the ability to exercise under optimal effort, and repeated intervention using rich feedback was effective in activating ankle joint function. In fact, before training with HAL-SJ, as shown in Figure 4, no obvious muscle activity was observed in either the left or right TA during ankle plantar dorsiflexion movement; however, after HAL-SJ intervention, muscle activity was observed in the left TA. The results of the 10-trial average showed phasic muscle activity during dorsiflexion in the left TA, suggesting an effect of training. It is not clear why muscle activity appeared only in the left TA despite interventions in both lower extremities, and the results of this study do not provide a complete explanation. In a report by Shimizu et al., HAL-SJ successfully induced the movement of paralyzed target muscles in a patient with cervical cord injury through performing of repetitive movements triggered by voluntary movements of other muscles [13]. However, in the present case, as the HAL-SJ was able to drive by the BES of the TA relatively early in the intervention; we speculate that the left lower extremity in this case had a potential for muscle contraction that was masked by the severe motor–sensory deficit. We believe that the advantage of using this device is that rich visual and motor feedback was provided to such masked potentials.

This study had some limitations. In this case, the HAL-SJ intervention was performed at approximately 100 days after injury. As this is a period of spontaneous recovery due to neuroplasticity after SCI [23], it may be difficult to completely determine the effect of HAL-SJ intervention. Therefore, it is not clear whether the effect of EMG appearance in the left TA obtained in this study was attributed to neuroplasticity or the activation and manifestation of voluntary movements that were difficult to perform due to severe motor sensory impairment. However, even after considering this issue, the appearance of EMG in the trained area was significant, and further improvements could have been achieved with prolonged training.

It should be noted that the device is subject to rental fees, which could be a potential cost and may make it difficult to use in all hospitals and institutions. Although the effects obtained with HAL-SJ alone were partial in this case, they may lead to a step up to another interventions using muscle activity. Thus, this device may be effective as one of the interventions that can be selected based on the patient's condition. In addition, ankle joint function is closely related to standing balance [24], and improvement of ankle joint function can lead to more practical gait by improving balance during walking [25], and may have a spillover effect on ADL, such as allowing selection of a simpler lower limb orthosis in the future. In the future, the number and frequency of effective interventions should be examined. Moreover, the extent to which these interventions are effective should be further verified.

4. Conclusions

We experienced and described a case in which intervention using HAL-SJ induced muscle potentials in a patient with a spinal cord injury who was unable to perform voluntary ankle movements due to severe motor–sensory dysfunction. Although intervention using the HAL-SJ could contribute to inducing muscle activity in this patient after SCI, further validation is needed to prove that HAL-SJ intervention induces locomotion in patients after SCI.

Author Contributions: Conceptualization, K.K., K.T. and H.M.; methodology, K.K., K.T., S.Y. and K.Y.; validation, K.K. and H.M.; formal analysis, K.K. and S.Y.; investigation, K.K., K.T. and A.A.; resources, A.A. and H.M.; data curation, K.K. and S.Y.; writing—original draft preparation, K.K.; writing—review and editing, K.K., K.T., S.Y., K.Y., A.A. and H.M.; visualization, K.K.; supervision, H.M.; project administration, H.M. All authors have read and agreed to the published version of the manuscript.

Funding: This research was founded by a Grant-in-Aid for Project Research (2269) from Ibaraki Prefectural University of Health Sciences.

Institutional Review Board Statement: The study was conducted in accordance with the Declaration of Helsinki, and the protocol was approved by the Ibaraki Prefectural University of Health Sciences Ethics Committee (approval number: 797; approval update and number: e356).

Informed Consent Statement: Sufficient explanations regarding the study procedures were provided to the patient before obtaining written consent for participation in the study. Informed consent was obtained from the patient.

Data Availability Statement: The data are contained within the article.

Acknowledgments: We would like to thank the Physical Therapy Department at Ibaraki Prefectural University of Health Sciences Hospital, represented by Yasutsugu Asakawa, the manager of the Department of Rehabilitation; Tomoyuki Matsuda, the manager of the Department of Physical Therapy; and Keiko Miyamoto, a robotic technical assistant.

Conflicts of Interest: The authors declare no conflict of interest.

References

1. Lo, J.; Chan, L.; Flynn, S. A systematic review of the incidence, prevalence, costs, and activity and work limitations of amputation, osteoarthritis, rheumatoid arthritis, back pain, multiple sclerosis, spinal cord injury, stroke, and traumatic brain injury in the United States: A 2019 Update. *Arch. Phys. Med. Rehabil.* **2021**, *102*, 115–131. [CrossRef]
2. Jain, N.B.; Ayers, G.D.; Peterson, E.N.; Harris, M.B.; Morse, L.; O'Connor, K.C.; Garshick, E. Traumatic spinal cord injury in the United States, 1993-2012. *JAMA* **2015**, *313*, 2236–2243. [CrossRef]
3. Miyakoshi, N.; Suda, K.; Kudo, D.; Sakai, H.; Nakagawa, Y.; Mikami, Y.; Suzuki, S.; Tokioka, T.; Tokuhiro, A.; Takei, H.; et al. A nationwide survey on the incidence and characteristics of traumatic spinal cord injury in Japan in 2018. *Spinal Cord* **2021**, *59*, 626–634. [CrossRef] [PubMed]
4. Ahuja, C.S.; Wilson, J.R.; Nori, S.; Kotter, M.R.N.; Druschel, C.; Curt, A.; Fehlings, M.G. Traumatic spinal cord injury. *Nat. Rev. Dis. Prim.* **2017**, *3*, 17018. [CrossRef] [PubMed]
5. Nas, K.; Yazmalar, L.; Şah, V.; Aydın, A.; Öneş, K. Rehabilitation of spinal cord injuries. *World J. Orthop.* **2015**, *6*, 8–16. [CrossRef] [PubMed]
6. Holanda, L.J.; Silva, P.M.M.; Amorim, T.C.; Lacerda, M.O.; Simão, C.R.; Morya, E. Robotic assisted gait as a tool for rehabilitation of individuals with spinal cord injury: A systematic review. *J. Neuroeng. Rehabil.* **2017**, *14*, 126. [CrossRef]
7. Mekki, M.; Delgado, A.D.; Fry, A.; Putrino, D.; Huang, V. Robotic rehabilitation and spinal cord injury: A narrative review. *Neurotherapeutics* **2018**, *15*, 604–617. [CrossRef]
8. Saita, K.; Morishita, T.; Hyakutake, K.; Fukuda, H.; Shiota, E.; Sankai, Y.; Inoue, T. Combined therapy using botulinum toxin A and single-joint hybrid assistive limb for upper-limb disability due to spastic hemiplegia. *J. Neurol. Sci.* **2017**, *373*, 182–187. [CrossRef]
9. Goto, K.; Morishita, T.; Kamada, S.; Saita, K.; Fukuda, H.; Shiota, E.; Sankai, Y.; Inoue, T. Feasibility of rehabilitation using the single-joint hybrid assistive limb to facilitate early recovery following total knee arthroplasty: A pilot study. *Assist. Technol.* **2017**, *29*, 197–201. [CrossRef]
10. Yoshikawa, K.; Koseki, K.; Endo, Y.; Yamamoto, S.; Kanae, K.; Takeuchi, R.; Yozu, A.; Mutsuzaki, H. Adjusting assistance commensurates with patient effort during robot-assisted upper limb training for a patient with spasticity after cervical spinal cord injury: A case report. *Medicina* **2019**, *55*, 404. [CrossRef] [PubMed]
11. Yoshioka, T.; Kubota, S.; Sugaya, H.; Arai, N.; Hyodo, K.; Kanamori, A.; Yamazaki, M. Feasibility and efficacy of knee extension training using a single-joint hybrid assistive limb, versus conventional rehabilitation during the early postoperative period after total knee arthroplasty. *J. Rural Med.* **2021**, *16*, 22–28. [CrossRef] [PubMed]
12. Takahashi, K.; Mutsuzaki, H.; Yoshikawa, K.; Yamamoto, S.; Koseki, K.; Takeuchi, R.; Mataki, Y.; Iwasaki, N. Robot-assisted ankle rehabilitation using the hybrid assistive limb for children after equinus surgery: A report of two cases. *Pediatr. Rep.* **2022**, *14*, 41. [CrossRef]
13. Shimizu, Y.; Kadone, H.; Kubota, S.; Ikumi, A.; Abe, T.; Marushima, A.; Ueno, T.; Endo, A.; Kawamoto, H.; Saotome, K.; et al. Active elbow flexion is possible in C4 quadriplegia using hybrid assistive limb (HAL®) technology: A case study. *J. Spinal Cord Med.* **2017**, *40*, 456–462. [CrossRef] [PubMed]
14. ASIA and ISCoS International Standards Committee. The 2019 revision of the International Standards for Neurological Classification of Spinal Cord Injury (ISNCSCI)-What's new? *Spinal Cord* **2019**, *57*, 815–817. [CrossRef]
15. Dietz, V.; Fouad, K. Restoration of sensorimotor functions after spinal cord injury. *Brain* **2014**, *137*, 654–667. [CrossRef]
16. Govil, K.; Noohu, M.M. Effect of EMG biofeedback training of gluteus maximus muscle on gait parameters in incomplete spinal cord injury. *NeuroRehabilitation* **2013**, *33*, 147–152. [CrossRef]

17. Roosink, M.; Mercier, C. Virtual feedback for motor and pain rehabilitation after spinal cord injury. *Spinal Cord* **2014**, *52*, 860–866. [CrossRef]
18. Cheung, E.Y.Y.; Yu, K.K.K.; Kwan, R.L.C.; Ng, C.K.M.; Chau, R.M.W.; Cheing, G.L.Y. Effect of EMG-biofeedback robotic-assisted body weight supported treadmill training on walking ability and cardiopulmonary function on people with subacute spinal cord injuries—A randomized controlled trial. *BMC Neurol.* **2019**, *19*, 140. [CrossRef]
19. Guo, Y.; Gao, F.; Li, J.; Yang, M.; Li, J.; Yang, D.; Du, L. Effect of electromyographic biofeedback training on motor function of quadriceps femoris in patients with incomplete spinal cord injury: A randomized controlled trial. *NeuroRehabilitation* **2021**, *48*, 345–351. [CrossRef] [PubMed]
20. Dietz, V. Body weight supported gait training: From laboratory to clinical setting. *Brain Res. Bull.* **2008**, *76*, 459–463. [CrossRef]
21. Benito-Penalva, J.; Edwards, D.J.; Opisso, E.; Cortes, M.; Lopez-Blazquez, R.; Murillo, N.; Costa, U.; Tormos, J.M.; Vidal-Samso, J.; Valls-Solé, J.; et al. Gait training in human spinal cord injury using electromechanical systems: Effect of device type and patient characteristics. *Arch. Phys. Med. Rehabil.* **2012**, *93*, 404–412. [CrossRef] [PubMed]
22. Field-Fote, E.C.; Roach, K.E. Influence of a locomotor training approach on walking speed and distance in people with chronic spinal cord injury: A randomized clinical trial. *Phys. Ther.* **2011**, *91*, 48–60. [CrossRef]
23. Burns, A.S.; Ditunno, J.F. Establishing prognosis and maximizing functional outcomes after spinal cord injury: A review of current and future directions in rehabilitation management. *Spine* **2001**, *26*, S137–S145. [CrossRef]
24. Lee, J.W.; Chan, K.; Unger, J.; Yoo, J.; Musselman, K.E.; Masani, K. Interjoint coordination between the ankle and hip joints during quiet standing in individuals with motor incomplete spinal cord injury. *J. Neurophysiol.* **2021**, *125*, 1681–1689. [CrossRef] [PubMed]
25. Calabrò, R.S.; Billeri, L.; Ciappina, F.; Balletta, T.; Porcari, B.; Cannavo, A.; Pignolo, L.; Manuli, A.; Naro, A. Toward improving functional recovery in spinal cord injury using robotics: A pilot study focusing on ankle rehabilitation. *Expert Rev. Med. Devices* **2022**, *19*, 83–95. [CrossRef] [PubMed]

Journal of
Functional Morphology and Kinesiology

MDPI

Article

EMG Amplitude–Force Relationship of Lumbar Back Muscles during Isometric Submaximal Tasks in Healthy Inactive, Endurance and Strength-Trained Subjects

Tim Schönau and Christoph Anders *

Division of Motor Research, Pathophysiology and Biomechanics, Experimental Trauma Surgery, Department for Hand, Reconstructive, and Trauma Surgery, Jena University Hospital, Friedrich-Schiller University, 07743 Jena, Germany
* Correspondence: christoph.anders@med.uni-jena.de; Tel.: +49-3641-9328-960

Abstract: Previous data suggest a correlation between the cross-sectional area of Type II muscle fibers and the degree of non-linearity of the EMG amplitude–force relationship (AFR). In this study we investigated whether the AFR of back muscles could be altered systematically by using different training modalities. We investigated 38 healthy male subjects (aged 19–31 years) who regularly performed either strength or endurance training (ST and ET, n = 13 each) or were physically inactive (controls (C), n = 12). Graded submaximal forces on the back were applied by defined forward tilts in a full-body training device. Surface EMG was measured utilizing a monopolar 4 × 4 quadratic electrode scheme in the lower back area. The polynomial AFR slopes were determined. Between-group tests revealed significant differences for ET vs. ST and C vs. ST comparisons at the medial and caudal electrode positions, but not for ET vs. C. Further, systematic main effects of the "electrode position" could be proven for ET and C groups with decreasing x^2 coefficients from cranial to caudal and lateral to medial. For ST, there was no systematic main effect of the "electrode position". The results point towards training-related changes to the fiber-type composition of muscles in the strength-trained participants, particularly for their paravertebral region.

Keywords: trunk muscles; training modality; amplitude–force relationship; human

Citation: Schönau, T.; Anders, C. EMG Amplitude–Force Relationship of Lumbar Back Muscles during Isometric Submaximal Tasks in Healthy Inactive, Endurance and Strength-Trained Subjects. *J. Funct. Morphol. Kinesiol.* **2023**, *8*, 29. https://doi.org/10.3390/jfmk8010029

Academic Editor: Giuseppe Musumeci

Received: 23 January 2023
Revised: 22 February 2023
Accepted: 22 February 2023
Published: 23 February 2023

1. Introduction

The amplitude–force relationship (AFR) during isometric contractions represents the connection between surface electromyography (SEMG) amplitude and functional muscle state, i.e., the muscle force expended. SEMG is the most common approach to evaluate muscle activity/excitation in vivo. Many factors, such as fiber-type composition, contraction velocity and muscle length, are determinants of the mechanical muscle output and thus the measured SEMG signal. Therefore the functional aspects of muscles, such as fatigue [1,2], muscle co-ordination [3–5], or muscle recruitment [6–8], can be quantitatively reflected by SEMG.

The systematic correlative relationship between the measured SEMG amplitude and the respective muscle force was described by Lippold in 1952 [9], followed by Bernshtein [10], Bouisset [11] and others. Back then, a linear AFR was assumed. To our knowledge, the first systematic analysis of the AFR concerning different muscles and differently trained subjects dates back to the early 1980s, when Lawrence and De Luca identified different AFR slopes [12]. They described specific characteristics that could be mathematically determined with either linear or polynomial regressions. These regressions were muscle-specific but did not show systematic alterations by activity level or type of sport. Subsequently, Solomonow and colleagues [13] performed stimulation experiments where they found a linear AFR in a fully recruited muscle; any further force increase was only due to an increase in the firing rate. A non-linear AFR was seen if an ongoing recruitment of motor units occurred together with a firing rate increase that accounted for the force increase. They, therefore, issued "a clear warning against a direct use of SEMG to predict muscle forces" [13].

If viewed from a morphological basis, the previous studies could already demonstrate different fiber-type compositions between muscles [14,15]. Functional priorities of certain muscles have further impact on the functional cross-sectional areas (CSA) of the two main fiber types which correlated nicely with endurance and a large CSA of Type I fibers and high power output and Type II fiber's CSA [14]. In addition, the adaption of functional fiber-type areas by specifically targeted training could be proven [16]. Our own previous investigations could confirm the known differences in the fiber-type composition of trunk muscles [17–19] by indirect measurements of the AFR. These were able to verify a muscle specificity with an either linear or non-linear AFR in trunk muscles with a non-linear AFR in the abdominal muscles and an almost-perfect linear AFR in the back muscles [20,21], extended by the fact that within this framework gender-specific AFR patterns could be identified [22]. Combining the facts of different fiber-type distributions in abdominal and back muscles, with a much larger CSA of Type I fibers in the back muscles [18] and a less pronounced non-linearity in women [22], the characteristics of AFR curves may be influenced by fiber-type composition. While the composition of fiber types is genetically predetermined and influenced by environmental factors [23], different training modalities alter muscle fiber types differently concerning their functional CSA.

Therefore, the current study aimed to identify the possible systematic alterations of the back muscle's AFR slopes by using different training modalities, which may alter the functional CSA of the two main fiber types. This approach has the potential to enhance the understanding of the back muscle's morpho-functional composition with respect to the different fiber-type areas. It therefore enables non-invasive diagnostics but additionally enables the tracing of training-, age-, or otherwise related muscular changes by electromyography. We recruited healthy volunteers who practiced either endurance or strength training at a competition-level and compared back muscle SEMG AFR curves between both groups and with a group of physically inactive participants. We expected a linear AFR in the endurance-trained and the inactive participants, whereas the strength-trained subjects were expected to show a non-linear AFR during graded defined submaximal isometric back muscle contraction tasks. As we applied a large quadratic electrode grid, independent of the training group, regional differences were expected because the investigated area included both paravertebral and more distant muscles, such as the latissimus dorsi and iliocostalis muscles. The latter ones are classified as more power-related, mobilizing muscles [24], whereas the paravertebral muscles are classified as stabilizing muscles [24,25]. Based on this and independent of the groups, we expected a lateral to medial change in AFR slopes from non-linear to a more linear characteristic.

2. Methods

2.1. Participants

For this study, 38 healthy male subjects aged between 19 and 31 years were recruited. The study cohort consisted of three groups based on their physical activity level (active and inactive) and type of training (see below). After being given information about the procedure and aim of the study, the subjects participated voluntarily and signed informed consent. The study was positively evaluated by the ethics committee of the Friedrich Schiller University Jena (2020-1844-BO).

The inclusion criteria for the two physically active groups were either endurance training (cycling or triathlon; endurance training group, ET, n = 13) or strength training (powerlifting or bodybuilding; strength training group, ST, n = 13) at a competition level (at least four training sessions per week, with a training duration of four to fifteen years). Participants who did both strength training and endurance training were not included. The subjects of the control group (C, n = 12) were not physically active at all (walking or participating in comparable activities once a week at most). The subjects' health status was briefly checked through medical history and physical examination. Besides general health problems interfering with study participation, specific exclusion criteria were acute or recurrent back pain and deformities or surgeries of the spine. A body height outside the range of 150 to 195 cm also led to exclusion, as the subjects would not have fit into the test and training device.

2.2. Device and Investigation

The tasks were performed in a computerized full body tilt device for the testing and training of trunk muscles (CTT Centaur, BfMC, Leipzig, Germany), which is used in numerous studies on trunk muscles [26–28]. In the device, the subjects were standing upright with their hips and legs fixed, while the upper body remained unsupported. To minimize the effects of varying upper extremity positions on the investigated trunk muscles, the participants held their arms crossed in front of their chests. In the device, the subjects could be tilted from a neutral upright to a horizontal position. During the tasks, the participants simply had to stabilize their upper body along their body axis while the device was tilted at defined angles between 0°and 90°. Control of proper upright upper body position was provided through a small biofeedback monitor positioned in front of the subjects. It contained a crosshair and a moving target point that deviated from the neutral position if any net force acted on the harness, positioned over the subject's shoulders. For this, the harness was equipped with strain gauges measuring forces in sagittal and frontal directions.

Defined forward tilt angles (0°, 5°, 10°, 20°, 30°, 45°, 60°, and 90°) were applied. Tilt angles could be converted into relative torque levels by applying the sine function and therefore corresponded to portions of 0%, 9%, 17%, 34%, 50%, 71%, 87%, and 100% of every subject's upper body weight (Figure 1). Tilt angles were randomly applied for ten seconds each to avoid order-dependent effects.

Figure 1. Applied tilt angles for the determination of the EMG amplitude–force relationship. Note that the upper body is always aligned along the body axis and the arms are held crossed in front of the chest.

2.3. Surface EMG Measurement and Analysis

Overall, 32 SEMG electrodes were applied in the lower back region on both sides utilizing a monopolar montage. For each side, a 4 × 4 quadratic electrode application scheme was used, to determine the position of the electrode (electrode position). This scheme was adjusted to each individual's anthropometry by defining the individual L1-L4 distance as the edge length (Figure 2). The determined L1-L4 distances varied between seven and nine centimeters. Each electrode arrangement was positioned 1.5 cm laterally from the midline, using prepared electrode grids. We used a quadratic electrode grid to avoid geometric distortions due to differences in individual anthropometry. Consequently, adhering to the quadratic geometry adjustment to the lumbar vertebral distances also resulted in a varying lateral dilation of the grid, including additional muscles of the back. With the lateral border of the grid ranging from 8.5 to 10.5 cm from the midline, specifically the lateral branches of the erector spinae (i.e., iliocostalis) and latissimus dorsi muscle activations were measured by the more lateral electrodes.

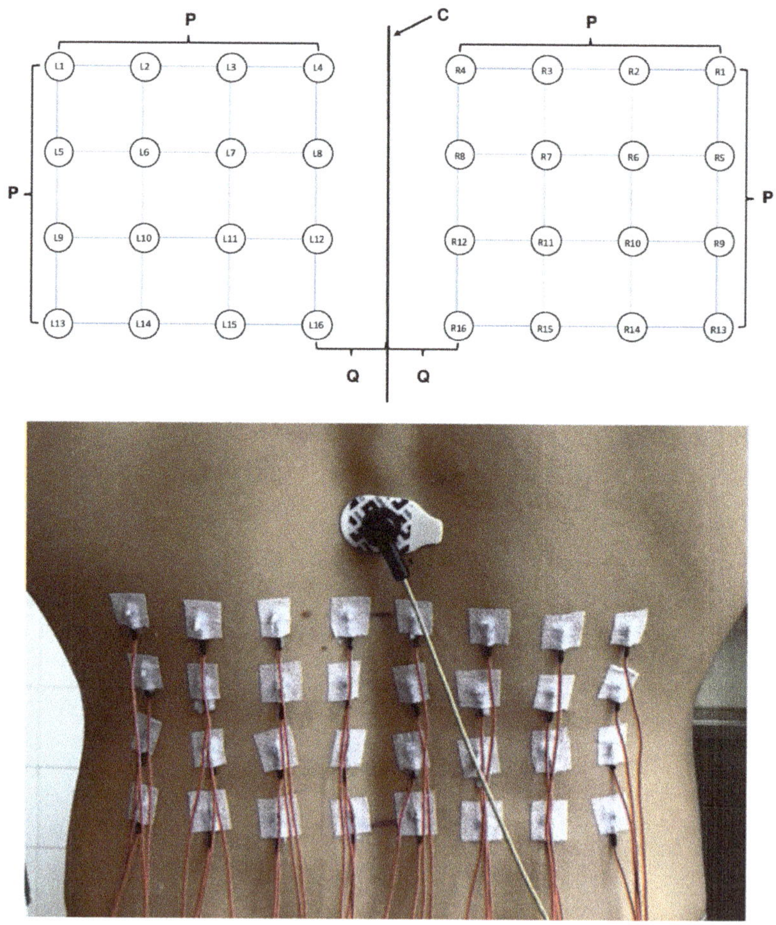

Figure 2. Top: Schematic display of the electrode montage. C: center line, P: edge length (adjusted to individual L1-L4 distance), Q: fixed distance of 1.5 cm from the center line. Electrode positions are indicated (L for left-sided, R for right-sided positions). Please note that electrode numbers are mirrored with respect to their lateral to medial orientation. Bottom: Electrode application as it appears on a participant.

For this investigation, reusable Ag-AgCl-electrodes with a diameter of six millimeters (DAGS102606, gvb geliMED, Bad Segeberg, Germany) and a hole in the middle of each electrode for electrode gel application (Electrode Cream, Care Fusion, Finland) were used. These electrodes were connected to a monopolar amplifier (ToEM16G: 10–1861 Hz (−3 dB), gain: 100, input impedance: 22 MOhm, SNR: 1.13 μV_{eff}, CMRR: 91.6 dB; DeMeTec, Germany). Additionally, two conjoined electrodes (disposable Ag-AgCl electrodes, H93SG, Covidien, Neustadt an der Donau, Germany) were attached over the subjects' anterior superior iliac spines, serving as reference electrodes whereas the ground electrode was attached at the Th11 level just over the palpable spinal process (compare with Figure 2). Electrocardiographic activation was recorded by the application of an additional bipolar input channel along the heart axis. All electrodes were applied after shaving and rubbing the examined regions with abrasive paste (SkinPure, Nihon Kohden, Tokyo, Japan). Electrodes were fixed with adhesive non-woven fabric. Special attention was paid to ensure caudal orientation of the electrode cables to avoid dislocation or levering off of the electrodes during task performance (compare with Figure 2).

Analog to digital conversion of the SEMG signals was performed with a sampling rate of 2048 per second (Tower of Measurement, DeMeTec, Langgöns, Germany, amplitude resolution: 24 bit at ± 5 V (6 nV/bit), anti-aliasing filter at 1024 Hz). SEMG data were stored on a computer (ATISArec, GJB, Ilmenau, Germany) for further analysis.

Data processing included the application of a high-pass filter at 20 Hz, a low-pass filter at 250 Hz and a notch filter at 50 Hz. ECG interferences were eliminated by using a template-based algorithm [29]. SEMG amplitude values were quantified as root mean square values, separately for each electrode and task.

As the AFR with respect to each group was the main outcome parameter, all AFR slopes were fitted applying a second-order polynomial function individually. The respective x^2 coefficients were then calculated for every electrode position. To improve numeral display and therefore the understanding of the differences between x^2 coefficients, all values were multiplied by 100. This accounts for all presented values.

2.4. Statistical Analysis

Statistical analysis of the calculated x^2 coefficients was performed by using IBM® SPSS® Statistics 28 (SPSS Inc., IBM, Corp., Armonk, NY, USA). A linear mixed-effects model (LMM) was fitted to compare the effects of "group", "side", and "electrode position" together with interactions between these factors (required significance level $p < 0.05$). In this analysis, "group", "side", and "electrode position" were modelled as fixed effects with a random intercept per subject. Initially, all interactions were calculated, but for the final analysis, only the significant interactions remained in the calculation.

3. Results

The initial LMM calculation of the x^2 coefficients "side" and "electrode position" together with the interaction of "group" * "electrode position" showed significant effects (all: $p < 0.001$).

As no other interactions showed significant results in addition to "side" and "electrode position"-related effects, group-specific differences that vary by electrode position were thus determined which were independent of their side. The explorative pairwise evaluation of group differences for the 16 side-independent electrode positions (now named EP as only their spatial arrangement was considered) showed that the x^2 coefficients were systematically different between groups at the medial and caudal positions for ET vs. ST and C vs. ST, but not for ET vs. C (Table 1). As can be taken from Table 1 irrespective of proven significant differences x^2 coefficients were always larger in the ST group.

For the main effect "side", x^2 coefficients were slightly larger for the right side ($p < 0.01$; left side 5.105, right side: 5.636). This was independent of the group. In both, the ET and C group x^2 coefficients always showed decreasing values from cranial to caudal and from lateral to medial positions (Figures 3 and 4). Systematic regional differences of the x^2 coefficients were detected for the ET and C groups (always $p < 0.001$, Figure 5) but not for ST ($p = 0.322$).

Table 1. Pairwise comparisons of the x^2 coefficients between groups per side-independent electrode position (EP).

	ET vs. ST			C vs. ST			ET vs. C		
		95% CI			95% CI			95% CI	
	Mean Diff.	Lower Border	Upper Border	Mean Diff.	Lower Border	Upper Border	Mean Diff.	Lower Border	Upper Border
EP1	−0.580	−4.388	3.228	−0.652	−4.629	3.326	0.072	−3.906	4.049
EP2	−1.168	−4.976	2.641	−1.304	−5.282	2.673	0.136	−3.841	4.114
EP3	−3.543	−7.351	0.265	−2.176	−6.154	1.801	−1.366	−5.344	2.611
EP4	−3.611	−7.419	0.197	−3.305	−7.283	0.672	−0.306	−4.283	3.672
EP5	−0.683	−4.491	3.126	−0.706	−4.683	3.272	0.023	−3.954	4.001
EP6	−1.598	−5.406	2.210	−1.106	−5.083	2.872	−0.493	−4.470	3.485
EP7	−3.546	−7.354	0.262	−3.347	−7.325	0.630	−0.199	−4.176	3.779
EP8	−3.393	−7.201	0.416	−3.226	−7.204	0.751	−0.167	−4.144	3.811
EP9	−1.064	−4.872	2.744	−0.990	−4.968	2.987	−0.073	−4.051	3.904
EP10	−1.940	−5.748	1.868	−2.120	−6.098	1.857	0.181	−3.797	4.158
EP11	−3.807	−7.615	0.001	−4.061	−8.039	−0.084	0.254	−3.723	4.232
EP12	−4.373	−8.182	−0.565	−4.318	−8.295	−0.340	−0056	−4.033	3.922
EP13	−1.286	−5.094	2.522	−1.465	−5.443	2.512	0.179	−3.798	4.157
EP14	−2.405	−6.213	1.403	−2.922	−6.899	1.056	0.517	−3.461	4.494
EP15	−3.823	−7.631	−0.015	−5.205	−9.183	−1.228	1.382	−2.595	5.360
EP16	−4.341	−8.150	−0.533	−4.547	−8.524	−0.569	0.206	−3.772	4.183

negative values: first group < second group. Shaded cells: $p < 0.05$ (adjustment for multiple tests: least significant difference).

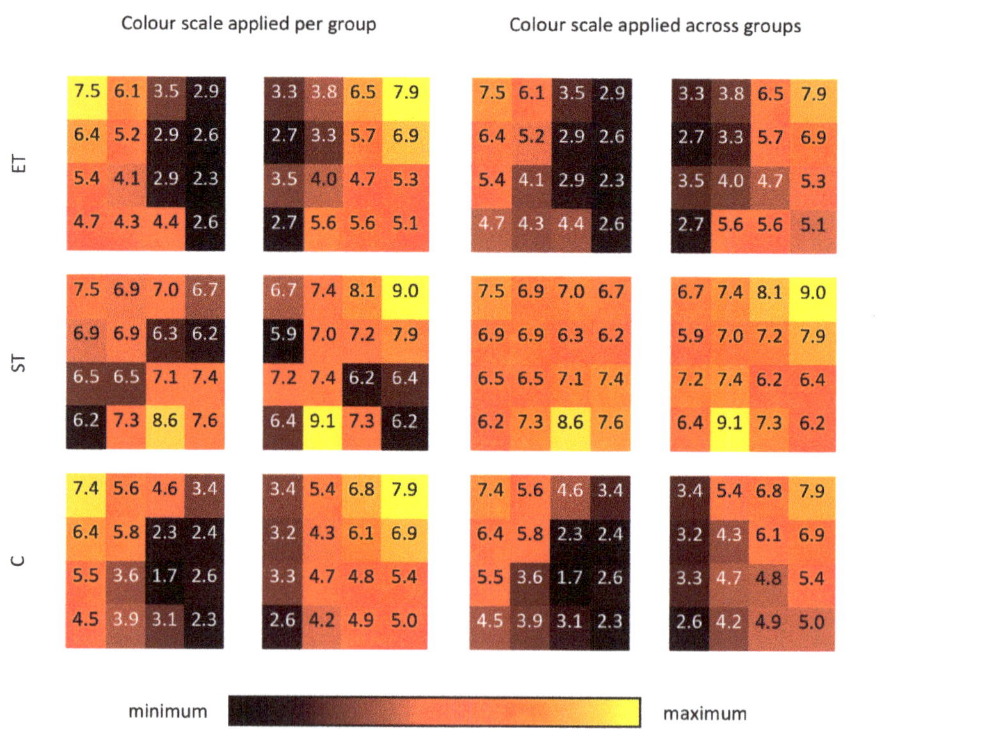

Figure 3. Map-like representation of mean x^2 coefficients per electrode position, separately for each group (note that values were multiplied by 100 for improved numeral display). The minimum to maximum color scale is either separately applied for each group (**left side**) or together for all groups (**right side**).

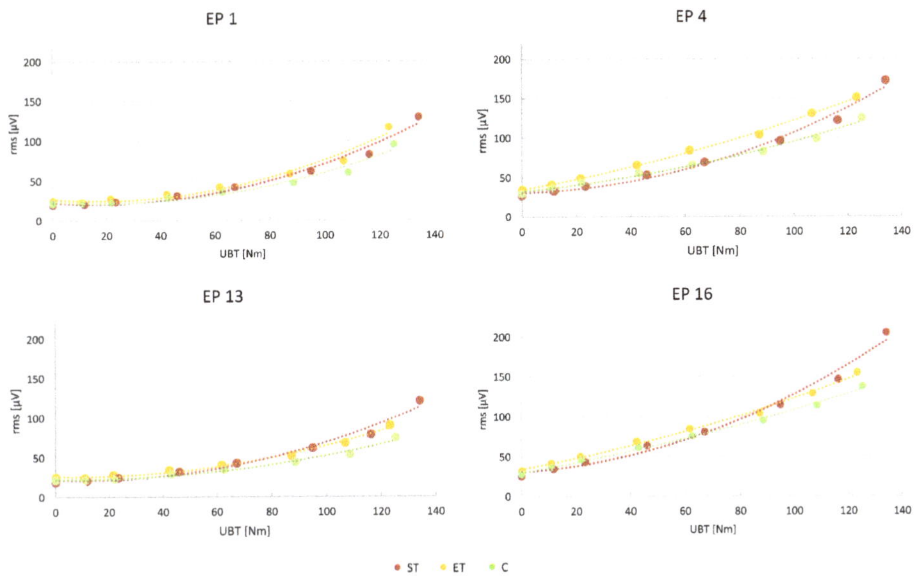

Figure 4. Examples of mean AFR slopes per group at selected positions at the left side. UBT: upper body torque.

ET	EP1	EP2	EP3	EP4	EP5	EP6	EP7	EP8	EP9	EP10	EP11	EP12	EP13	EP14	EP15	EP16
EP1		1.396	4.065	4.621	1.021	2.253	4.586	5.042	2.340	3.307	4.255	4.797	2.820	2.803	2.675	5.044
EP2			2.669	3.225	-0.374	0.857	3.190	3.646	0.944	1.911	2.860	3.402	1.424	1.407	1.279	3.648
EP3				0.556	-3.044	-1.812	0.521	0.977	-1.725	-0.758	0.190	0.732	-1.245	-1.262	-1.390	0.979
EP4					-3.599	-2.368	-0.035	0.421	-2.281	-1.314	-0.365	0.177	-1.800	-1.818	-1.946	0.423
EP5						1.232	3.565	4.021	1.319	2.285	3.234	3.776	1.799	1.781	1.654	4.023
EP6							2.333	2.789	0.087	1.054	2.002	2.544	0.567	0.550	0.422	2.791
EP7								0.456	-2.246	-1.279	-0.331	0.211	-1.766	-1.783	-1.911	0.458
EP8									-2.702	-1.735	-0.787	-0.245	-2.222	-2.239	-2.367	0.002
EP9										0.967	1.915	2.457	0.480	0.463	0.335	2.704
EP10											0.949	1.491	-0.486	-0.504	-0.632	1.737
EP11												0.542	-1.435	-1.453	-1.580	0.789
EP12													-1.977	-1.995	-2.122	0.247
EP13														-0.017	-0.145	2.224
EP14															-0.128	2.241
EP15																2.369
EP16																

C	EP1	EP2	EP3	EP4	EP5	EP6	EP7	EP8	EP9	EP10	EP11	EP12	EP13	EP14	EP15	EP16
EP1		1.461	2.627	4.243	0.973	1.689	4.316	4.804	2.195	3.416	4.438	4.670	2.928	3.248	3.986	5.178
EP2			1.167	2.783	-0.488	0.228	2.855	3.343	0.734	1.955	2.977	3.209	1.467	1.787	2.525	3.717
EP3				1.616	-1.654	-0.938	1.688	2.177	-0.432	0.788	1.811	2.043	0.301	0.620	1.359	2.551
EP4					-3.270	-2.554	0.072	0.560	-2.049	-0.828	0.195	0.427	-1.315	-0.996	-0.258	0.934
EP5						0.716	3.342	3.831	1.222	2.442	3.465	3.697	1.955	2.275	3.013	4.205
EP6							2.627	3.115	0.506	1.727	2.749	2.981	1.239	1.559	2.297	3.489
EP7								0.488	-2.121	-0.900	0.122	0.355	-1.387	-1.068	-0.330	0.862
EP8									-2.609	-1.388	-0.366	-0.134	-1.876	-1.556	-0.818	0.374
EP9										1.221	2.243	2.475	0.733	1.053	1.791	2.983
EP10											1.022	1.254	-0.487	-0.168	0.570	1.762
EP11												0.232	-1.510	-1.190	-0.452	0.740
EP12													-1.742	-1.422	-0.684	0.508
EP13														0.320	1.058	2.250
EP14															0.738	1.930
EP15																1.192
EP16																

Figure 5. Pairwise comparisons of the x^2 coefficients for ET (**upper panel**) and C (**lower panel**) groups between side-independent electrode positions (EP). Values correspond to the mean differences of each comparison. Significant (least significant difference) differences between x^2 coefficient values are indicated by shadings. Brown shading: line heading position > column heading position; grey shading: line heading position < column heading position.

4. Discussion

In the present study, the amplitude–force relationship of the lower back muscles was investigated during submaximal load situations. To compare the curve characteristics of the respective AFR, second order polynomials were fitted. Their x^2 coefficients reflect the extent of non-linearity, i.e., large x^2 coefficients indicate strong non-linearity of the AFR, whereas low values stand for a more linear curve characteristic.

In general, a non-linear AFR for ST and linear AFR for ET and C was not statistically proven. Furthermore, the detection of a possible main effect "group" in the linear mixed model did not show a significant influence regarding the x^2 coefficients. On the other hand, in the LMM, a significant interaction was proven for "group" * "electrode position", meaning that the different training modalities affected the x^2 coefficients at several electrode positions. Further, in the ET and C groups systematic differences between the x^2 coefficients were proven concerning electrode position, where the degree of non-linearity dropped from cranial to caudal and lateral to medial. In contrast, such systematic spatial differences were not proven in the ST group.

4.1. Recruitment Strategies

The demonstrated position- and group-specific differences of the AFR may be explained by the different recruitment strategies of the investigated muscles [13]. Published data could show that the curve characteristics of the AFR depend on how the tested muscle achieves additional power output [13]. Thus, a non-linear AFR is evident with combined recruitment of additional motor units and an increase in firing rate [13]. Linear curve slopes were proven if mainly the firing rate of the active motor units increased [13]. However, these results were obtained under stimulation conditions. The respective recruitment strategy seems to be muscle-specific [30,31]. This muscle specificity is determined by fiber composition, function/task, and the number of motor units of the respective muscle.

4.2. Muscle Fiber Composition

The found position-specific differences may be explained by different muscle fiber compositions and muscle fiber diameters of the investigated muscles. Muscle fiber biopsies in healthy untrained men aged 20–30 years of the longissimus and multifidus muscles at the L3 level revealed a Type I fiber content of 57–65% for both studied muscles [18,19]. Thus, there was no difference between the medially located multifidus muscle and the more laterally located longissimus muscle. Post-mortem studies by Johnson et al. revealed an average of 58% Type I fibers for the erector spinae muscle [14]. It can be postulated that due to the high stability requirements, predominantly Type I fibers occur in the medial portions of the lumbar erector spinae muscle (longissimus, multifidus, spinal, etc.), compared to more laterally located portions (iliocostalis muscle). Unfortunately, we did not find any study that has specifically investigated the iliocostalis muscle's fiber composition. Since due to their anatomical location latissimus dorsi and iliocostalis muscles are both used less for stability but more for movement execution [32], an elevated Type II fiber content can be assumed. In the study by Johnson et al. a ratio of 50% Type I fibers could be demonstrated in the latissimus dorsi muscle [14]. The more heterogeneous the muscle fiber composition of a muscle and the larger the Type II muscle fiber CSA, the more non-linear the AFR will be [33]. Our own previous studies with untrained subjects showed a non-linear AFR for abdominal muscles, whereas an almost ideal linear AFR was found in the back muscles [20]. According to the literature, abdominal muscles show a Type I fiber content of 46–58% [14,17], whereas back muscles contain 57–65% Type I fibers [14,24,25]. This fact and the current results support the hypothesis of a fiber-type-dependent curve characteristic of the AFR. Type II muscle fibers exhibit a higher density of Na^+ channels, faster action potentials, and a higher resting membrane potential [33]. During isometric contractions, motor units are recruited in an ordered sequence: firstly, fibers with a low shortening velocity (Type I) and secondly, fibers with a higher shortening velocity (Type II) [34,35]. This results in disproportionately larger potentials at higher force levels, which are generated primarily by motor units of Type II fibers, producing an exponential and therefore non-linear AFR. This cannot be demonstrated in muscles with a predominant Type I portion [31]. In the current investigation, all subjects showed a non-linear AFR in the most lateral and cranial region of the investigated region, where mainly latissimus dorsi and iliocostalis muscles are to be found. The respective AFR can consequently be explained by the significantly more heterogeneous fiber composition of these muscles.

4.3. Training Associated Effects

The ST participants showed no group-specific systematic differences in x^2 coefficients concerning electrode position, but consistently large x^2 coefficients that were significantly different from the two other groups at the caudal and medial positions. Although the individual proportion of Type I and Type II fibers is genetically predetermined and influenced by environmental factors [23], fiber composition and CSA can be altered by training evoked adaptation [36]. Studies have shown that pure endurance training does not affect functional muscle fiber CSAs [37,38]. Strength training and combined endurance and strength training both increase Type II fibers CSAs by 13–18%. Isolated strength training additionally increased Type I fiber CSA [38]. Although we did not take muscle biopsies, the mentioned training-induced effects can also be assumed in our active groups because all of our trained subjects performed years of intense training in their sports. As already mentioned, systematic differences between groups could only be determined for the most caudal and medial positions. The more lateral electrode positions contain information of lateral branches of the erector spinae, i.e., iliocostalis and also latissimus dorsi muscles, which showed a clear non-linear AFR also in the ET and C groups and therefore precluded any systematic differences between the groups. It is reasonable to assume that the described training-associated effects on muscle fibers in the ST group caused areas with assumable linear AFR (medial and caudal) to change to a non-linear AFR. Medial and caudal linear AFRs were detected in the ET and C groups, but not in the ST group. Thus, the position-specific effects found in the ET and C groups were abolished by strength training, which may also result in the non-systematic x^2 coefficients distribution found in the ST group in the per-group analysis. This explains the found systematic differences in the group-electrode-interaction analysis in the group comparisons (Table 1). The described remaining non-linearity of the AFR in the medial muscle components of the ST participants ensured that group-specific differences were most pronounced in this area.

We found significant side effects with a higher non-linearity on the right side (compare with Figure 3). This may be explained by the fact that, except for two subjects (one ET, one C), only right-handed individuals participated in our study. The contralateral side of the back muscles is known to be more fatigue-resistant due to their predominant stability function when working with the right upper extremity [39]. Therefore, a higher proportion of Type I fibers compared to the ipsilateral side seems to be likely. This may result in an overall higher non-linearity on the right side with respect to the contralateral side.

4.4. Clinical Implications

Our trunk and particularly our back muscles play a key role in mobilizing and stabilizing the spine. Mainly impaired back muscle force control [40] or corrupted coordination causing subfailure injuries [41] are considered as relevant causes for back pain. In contrast, differences in maximum force production seem to be of inferior importance concerning the development of lower back pain [42]. In this respect, isolated strength training does not seem to be the decisive factor for the prevention of lower back pain. Moreover, if during sustained muscle activity, not only fatigue-related muscle force, but also co-ordination is corrupted [43], then acute episodes of lower back pain are more likely to occur. Therefore, the presented data are another strong argument for the application of functionally oriented physical training in back pain patients to prevent episodes of acute back pain and also to restore the necessary equilibrium between the stability and mobility of the spine. For this, multimodal rehabilitation programs were developed which not only aim to improve the physical state of the patients, but also add educative parts in order to restore or even built more self-reliant movement patterns, based on regained physical possibilities without fear of exercise. With respect to the treatment of lower back pain, not only are functional parameters known to be important, but also psychological characteristics [44], but these were not the focus of the current investigation.

4.5. Limitations

As no muscle biopsies were taken, the drawn conclusions about the fiber composition remain somehow speculative. On the other hand, as could be proven elsewhere during an endurance test exploring the same population [45] the participants of the ST group were more prone to back muscle fatigue in comparison to the other two groups. This indirectly proves the correctness of the assumed differences of the functional fiber-type areas between the groups.

The applied submaximal static tests were performed in a specific test and training device, which is not available everywhere. Besides this drawback, using the device we could apply exactly defined loads of the mentioned portions from 9% to 100% upper body weight to all participants without any uncertainty regarding the correct measurement of each participant's upper body weight [28] by simply tilting them at the respective tilt angles in the sagittal plane.

The electrodes were applied using a monopolar montage, which is much more prone to cross-talk than bipolar montages [46]. We decided to use this montage, since therewith we could align the electrodes according to each individual's morphology. The natural drawback of this individual alignment was that inter-electrode distances vary between subjects, and therefore prohibit bipolar signal calculation.

5. Conclusions/Suggestions for Future Research

Isolated strength training is accompanied by relevant changes in the AFR of back muscles, especially for their paravertebral portions. These modifications can be attributed to sport-specific changes in fiber composition. As part of the evolutionary development of upright posture and the associated use of the arms for load manipulation, based on their fiber composition paravertebral muscles have become more specialized for postural tasks. Therefore, remodeling processes with an increased Type II fiber area could have rather detrimental consequences in the long term.

As the results showed clear differences between the groups, training or rehabilitation programs may be accompanied by SEMG measurements to monitor the induced changes in functional capacity. The effects of different training methods could thereby be evaluated. As only submaximal forces were applied, the results are mostly independent from the subject's motivation.

Author Contributions: Conceptualization, C.A.; methodology, C.A. and T.S.; software, C.A.; validation, C.A. and T.S.; formal analysis, T.S.; investigation, T.S.; resources, C.A.; data curation, C.A., T.S.; writing—original draft preparation, C.A. and T.S.; writing—review and editing, C.A. and T.S.; visualization, C.A. and T.S.; supervision, C.A.; project administration, C.A.; funding acquisition, C.A. All authors have read and agreed to the published version of the manuscript.

Funding: Laboratory facilities and research grants (2.11.11.20/21) were funded by the Centre of Interdisciplinary Prevention of Diseases related to Professional Activities (KIP) funded by Friedrich Schiller University Jena and the Berufsgenossenschaft Nahrungsmittel und Gastgewerbe. The sponsors were not involved in the study and publication progress.

Institutional Review Board Statement: The study was conducted in accordance with the Declaration of Helsinki and approved by the ethics committee of the Friedrich Schiller University Jena (2020-1844-BO).

Informed Consent Statement: Informed consent was obtained from all subjects involved in the study.

Data Availability Statement: All data are available upon request.

Acknowledgments: The authors wish to thank Elke Mey for technical assistance and Thomas Lehmann for help with the statistical analysis.

Conflicts of Interest: Laboratory facilities and research grants (2.11.11.20/21) were provided by the Centre of Interdisciplinary Prevention of Diseases related to Professional Activities (KIP) funded by Friedrich Schiller University Jena and the Berufsgenossenschaft Nahrungsmittel und Gastgewerbe. The sponsors were not involved in the study and publication progress.

References

1. Gandevia, S.C. Spinal and Supraspinal Factors in Human Muscle Fatigue. *Physiol. Rev.* **2001**, *81*, 1725–1789. [CrossRef] [PubMed]
2. Luttmann, A.; Jäger, M.; Laurig, W. Electromyographical indication of muscular fatigue in occupational field studies. *Int. J. Ind. Ergon.* **2000**, *25*, 645–660. [CrossRef]
3. Anders, C.; Wenzel, B. Activation characteristics of trunk muscles during cyclic upper body pertubations. *Acta Physiol.* **2006**, *186*, 245.
4. Graven-Nielsen, T.; Svensson, P.; Arendt-Nielsen, L. Effects of experimental muscle pain on muscle activity and co-ordination during static and dynamic motor function. *Electromyogr. Motor C* **1997**, *105*, 156–164. [CrossRef] [PubMed]
5. Van Dieën, J.H.; Cholewicki, J.; Radebold, A. Trunk Muscle Recruitment Patterns in Patients with Low Back Pain Enhance the Stability of the Lumbar Spine. *Spine* **2003**, *28*, 834–841. [CrossRef]
6. Chapman, A.R.; Vicenzino, B.; Blanch, P.; Hodges, P.W. Patterns of leg muscle recruitment vary between novice and highly trained cyclists. *J. Electromyogr. Kinesiol.* **2008**, *18*, 359–371. [CrossRef]
7. Fuglevand, A.J.; Winter, D.A.; Patla, A.E. Models of recruitment and rate coding organization in motor-unit pools. *J. Neurophysiol.* **1993**, *70*, 2470–2488. [CrossRef]
8. Urquhart, D.M.; Hodges, P.W.; Allen, T.J.; Story, I.H. Abdominal muscle recruitment during a range of voluntary exercises. *Man. Ther.* **2005**, *10*, 144–153. [CrossRef]
9. Lippold, O.C.J. The relation between integrated action potentials in a human muscle and its isometric tension. *J. Physiol.* **1952**, *117*, 492–499. [CrossRef]
10. Bernshtein, V.M. Statistical parameters of the electrical signal of a muscle model. *Biofizika* **1967**, *12*, 693–703.
11. Bouisset, S. EMG and Muscle Force in Normal Motor Activities. In *New Developments in Electromyography and Clinical Neurophysiology*; Desmedt, J.E., Ed.; Karger: Basel, Switzerland, 1973; Volume 1, pp. 547–583.
12. Lawrence, J.H.; De Luca, C.J. Myoelectric signal versus force relationship in different human muscles. *J. Appl. Physiol.* **1983**, *54*, 1653–1659. [CrossRef]
13. Solomonow, M.; Baratta, R.; Shoji, H.; D'Ambrosia, R. The EMG-force relationships of skeletal muscle; dependence on contraction rate, and motor units control strategy. *Electromyogr. Clin. Neurophysiol.* **1990**, *30*, 141–152.
14. Johnson, M.A.; Polgar, J.; Weightman, D.; Appleton, D. Data on the distribution of fibre types in thirty-six human muscles: An autopsy study. *J. Neurol. Sci.* **1973**, *18*, 111–129. [CrossRef]
15. Gollnick, P.D.; Armstrong, R.B.; Saubert, C.W.; Piehl, K.; Saltin, B. Enzyme activity and fiber composition in skeletal muscle of untrained and trained men. *J. Appl. Physiol.* **1972**, *33*, 312–319. [CrossRef]
16. Gollnick, P.D.; Armstrong, R.B.; Saltin, B.; Saubert, C.W.; Sembrowich, W.L.; Shepherd, R.E. Effect of training on enzyme activity and fiber composition of human skeletal muscle. *J. Appl. Physiol.* **1973**, *34*, 107–111. [CrossRef]
17. Häggmark, T.; Thorstensson, A. Fibre types in human abdominal muscles. *Acta Physiol. Scand.* **1979**, *107*, 319–325. [CrossRef]
18. Mannion, A.F. Fibre type characteristics and function of the human paraspinal muscles: Normal values and changes in association with low back pain. *J. Electromyogr. Kinesiol.* **1999**, *9*, 363–377. [CrossRef]
19. Thorstensson, A.; Carlson, H. Fibre types in human lumbar back muscles. *Acta Physiol. Scand.* **1987**, *131*, 195–202. [CrossRef]
20. Anders, C.; Brose, G.; Hofmann, G.O.; Scholle, H.-C. Evaluation of the EMG–force relationship of trunk muscles during whole body tilt. *J. Biomech.* **2008**, *41*, 333–339. [CrossRef]
21. Huebner, A.; Faenger, B.; Scholle, H.-C.; Anders, C. Re-evaluation of the amplitude–force relationship of trunk muscles. *J. Biomech.* **2015**, *48*, 1198–1205. [CrossRef]
22. Anders, C.; Brose, G.; Hofmann, G.O.; Scholle, H.-C. Gender specific activation patterns of trunk muscles during whole body tilt. *Eur. J. Appl. Physiol.* **2007**, *101*, 195–205. [CrossRef]
23. Simoneau, J.A.; Bouchard, C. Genetic determinism of fiber type proportion in human skeletal muscle. *FASEB J.* **1995**, *9*, 1091–1095. [CrossRef] [PubMed]
24. Comerford, M.J.; Mottram, S.L. Movement and stability dysfunction—contemporary developments. *Man. Ther.* **2001**, *6*, 15–26. [CrossRef] [PubMed]
25. Bergmark, A. Stability of the lumbar spine: A study in mechanical engineering. *Acta Orthop.* **1989**, *60*, 1–54. [CrossRef] [PubMed]
26. Hansen, L.; Anders, C. Influence of different control strategies on muscle activation patterns in trunk muscles. *Physiol. Rep.* **2014**, *2*, e12229. [CrossRef]
27. Anders, C.; Steiniger, B. Main force directions of trunk muscles: A pilot study in healthy male subjects. *Hum. Mov. Sci.* **2018**, *60*, 214–224. [CrossRef]

28. Pfeifle, C.; Edel, M.; Schleifenbaum, S.; Kühnapfel, A.; Heyde, C.-E. The reliability of a restraint sensor system for the computer-supported detection of spinal stabilizing muscle deficiencies. *BMC Musculoskelet. Disord.* **2020**, *21*, 1–9. [CrossRef]
29. Mörl, F.; Anders, C.; Grassme, R. An easy and robust method for ECG artifact elimination of SEMG signals. In Proceedings of the XVII Congress of the International Society of Electrophysiology and Kinesiology Aalborg, Aalborg, Denmark, 16–19 June 2010.
30. De Luca, C.J.; LeFever, R.S.; McCue, M.P.; Xenakis, A.P. Behaviour of human motor units in different muscles during linearly varying contractions. *J. Physiol.* **1982**, *329*, 113–128. [CrossRef]
31. Kukulka, C.G.; Clamann, H.P. Comparison of the recruitment and discharge properties of motor units in human brachial biceps and adductor pollicis during isometric contractions. *Brain Res.* **1981**, *219*, 45–55. [CrossRef]
32. Gibbons, S.G.T.; Comerford, M.J. Strength versus stability: Part 1: Concept and terms. *Orth. Div. Rev.* **2001**, *43*, 21–27.
33. Woods, J.J.; Bigland-Ritchie, B. Linear and non-linear surface EMG/force relationships in human muscles. An anatomical/functional argument for the existence of both. *Am. J. Phys. Med.* **1983**, *62*, 287–299.
34. Farina, D.; Fosci, M.; Merletti, R. Motor unit recruitment strategies investigated by surface EMG variables. *J. Appl. Physiol.* **2002**, *92*, 235–247. [CrossRef]
35. Milner-Brown, H.S.; Stein, R.B.; Yemm, R. The orderly recruitment of human motor units during voluntary isometric contractions. *J. Physiol.* **1973**, *230*, 359–370. [CrossRef]
36. Goldspink, G.; Scutt, A.; Loughna, P.T.; Wells, D.J.; Jaenicke, T.; Gerlach, G.F. Gene expression in skeletal muscle in response to stretch and force generation. *Am. J. Physiol.* **1992**, *262*, R356–R363. [CrossRef]
37. Kraemer, W.J.; Patton, J.F.; Gordon, S.E.; Harman, E.A.; Deschenes, M.R.; Reynolds, K.; Newton, R.U.; Triplett, N.T.; Dziados, J.E. Compatibility of high-intensity strength and endurance training on hormonal and skeletal muscle adaptations. *J. Appl. Physiol.* **1995**, *78*, 976–989. [CrossRef]
38. Putman, C.T.; Xu, X.; Gillies, E.; MacLean, I.M.; Bell, G.J. Effects of strength, endurance and combined training on myosin heavy chain content and fibre-type distribution in humans. *Eur. J. Appl. Physiol.* **2004**, *92*, 376–384. [CrossRef]
39. Merletti, R.; De Luca, C.J.; Sathyan, D. Electrically evoked myoelectric signals in back muscles: Effect of side dominance. *J. Appl. Physiol.* **1994**, *77*, 2104–2114. [CrossRef]
40. Pranata, A.; Perraton, L.; El-Ansary, D.; Clark, R.; Fortin, K.; Dettmann, T.; Brandham, R.; Bryant, A. Lumbar extensor muscle force control is associated with disability in people with chronic low back pain. *Clin. Biomech.* **2017**, *46*, 46–51. [CrossRef]
41. Panjabi, M.M. A hypothesis of chronic back pain: Ligament subfailure injuries lead to muscle control dysfunction. *Eur. Spine J.* **2005**, *15*, 668–676. [CrossRef]
42. Da Silva, R.A., Jr.; Arsenault, A.B.; Gravel, D.; Larivière, C.; de Oliveira, E., Jr. Back muscle strength and fatigue in healthy and chronic low back pain subjects: A comparative study of 3 assessment protocols. *Arch. Phys. Med. Rehabil.* **2005**, *86*, 722–729. [CrossRef]
43. Anders, C.; Kankaanpää, M.; Airaksinen, O.; Scholle, H.C.; Hänninen, O. Coordination of lumbar back muscles of dynamic loading. *Man. Med.* **1998**, *36*, 61–65. [CrossRef]
44. Hildebrandt, J.; Pfingsten, M.; Saur, P.; Jansen, J. Prediction of success from a multidisciplinary treatment pro-gram for chronic low back pain. *Spine* **1997**, *22*, 990–1001. [CrossRef] [PubMed]
45. Anders, C.; Schönau, T. Spatiotemporal characteristics of lower back muscle fatigue during a ten minutes endurance test at 50% upper body weight in healthy inactive, endurance, and strength trained subjects. *PLoS ONE* **2022**, *17*, e0273856. [CrossRef] [PubMed]
46. Mohr, M.; Schön, T.; Von Tscharner, V.; Nigg, B.M. Intermuscular Coherence Between Surface EMG Signals Is Higher for Monopolar Compared to Bipolar Electrode Configurations. *Front. Physiol.* **2018**, *9*, 566. [CrossRef]

Journal of
Functional Morphology and Kinesiology

MDPI

Article

Association of Strength Performance in Bench Press and Squat with Anthropometric Variables between Resistance-Trained Males and Females

Hallvard Nygaard Falch, Markus Estifanos Haugen, Stian Larsen and Roland van den Tillaar *

Department of Sports Sciences and Physical Education, Nord University, 7600 Levanger, Norway
* Correspondence: roland.v.tillaar@nord.no

Abstract: Individual differences in the appropriate percentage of 1-RM for a given repetition range could be a result of variation in anthropometrics and/or sex. Strength endurance is the term used to describe the ability to perform a number of repetitions prior to failure (AMRAP) in sub-maximal lifts and is important in determining the appropriate load for the targeted repetition range. Earlier research investigating the association of AMRAP performance and anthropometric variables was often performed in a sample of pooled sexes or one sex only or by utilizing tests with low ecological validity. As such, this randomized cross-over study investigates the association of anthropometrics with different measures of strength (maximal and relative strength and AMRAP) in the squat and bench press for resistance-trained males (n = 19, 24.3 ± 3.5 years, 182 ± 7.3 cm, 87.1 ± 13.3 kg) and females (n = 17, 22.1 ± 3 years, 166.1 ± 3.7 cm, 65.5 ± 5.6 kg) and whether the association differs between the sexes. Participants were tested for 1-RM strength and AMRAP performance, with 60% of 1-RM in the squat and bench press. Correlational analysis revealed that for all participants, lean mass and body height were associated with 1-RM strength in the squat and bench press (0.66, $p \leq 0.01$), while body height was inversely associated with AMRAP performance ($r \leq -0.36, p \leq 0.02$). Females had lower maximal and relative strength with a greater AMRAP performance. In the AMRAP squat, thigh length was inversely associated with performance in males, while fat percentage was inversely associated with performance in females. It was concluded that associations between strength performance and anthropometric variables differed for males and females in fat percentage, lean mass, and thigh length.

Keywords: strength training; sex; 1-RM performance

Citation: Falch, H.N.; Haugen, M.E.; Larsen, S.; van den Tillaar, R. Association of Strength Performance in Bench Press and Squat with Anthropometric Variables between Resistance-Trained Males and Females. *J. Funct. Morphol. Kinesiol.* **2023**, *8*, 19. https://doi.org/10.3390/jfmk8010019

Academic Editor: Diego Minciacchi

Received: 28 December 2022
Revised: 28 January 2023
Accepted: 29 January 2023
Published: 1 February 2023

1. Introduction

The benefits of resistance training are numerous, as incorporating habitual resistance training may increase muscular strength and muscular cross-sectional area, as well as improve markers of health, function in daily living [1–3], and sporting performance [4–7]. Enhancements in strength can be defined as an increased ability to exert force against external resistance [7], whereby strength again can be divided into several sub-categories, such as maximal strength (absolute force exertion), relative strength (force exerted per unit of body mass), and strength endurance (ability to resist fatigue and reductions in force output) [8]. Resistance training for improving sporting performance varies based on the distinctiveness and needs analysis for different sports. For example, in collision sports, there is a great focus on maximal strength (absolute force), despite increasing body mass, due to the importance of sprint momentum [9]. In team sports, however, where athletes have to accelerate their body mass, strength relative to their body mass seems like the more appropriate training goal [4,10,11]. In sports requiring multiple repetitions of a similar movement, such as Crossfit™, high levels of local muscular endurance are required of the athlete [12].

Males commonly possess higher levels of maximal and relative strength [13], mainly due to a greater amount of muscle mass and a lower body fat percentage [14,15], increasing the force capacity per unit of body mass. On the other hand, females are thought to outperform men in strength endurance tasks [8,15–17]. The sex-related differences observed in fatiguability when performing strength endurance work are not fully understood, but some mechanisms have been proposed. It could be a result of females possessing a greater proportion of fatigue-resistant type I fibers [15,18]. Sex differences are also studied in metabolism, suggesting males have a greater reliance on glycolytic pathways as opposed to females' greater reliance on fat oxidation [8]. Another proposed mechanism is that lower absolute muscle force at a similar relative intensity leads to less intramuscular compression, thus enhancing oxygen availability and allowing easier clearance of metabolites through greater blood flow. In conjunction, less muscle mass would also reduce oxygen requirements [8]. From a mechanical perspective, shorter segment lengths lead to less work (force x distance) and external torque requirements (load x moment arm) per repetition being performed [13], a mechanical advantage for females due to a commonly shorter stature and length of limbs. The resistive torque and work requirements can be manipulated within exercises, such as by adjusting grip width in the bench press [19], although longer limbs are a disadvantage in theory.

The extent to which sex-related strength differences are observed is furthermore suggested to be task-specific [17,20], as the sex-related difference in maximal strength has been indicated to be greater in the upper limbs compared to the lower limbs [21,22]. Additionally, strength endurance performance favoring females seems to be greater in isometric tests, with work being performed at a lower percentage of maximal voluntary contraction (<80% of 1-RM) [8,16]. The aforementioned considerations are important when prescribing training programs, as resistance training programs often prescribe loads to athletes based on percentages of their 1-RM [23] to elicit specific adaptations. However, the maximal number of repetitions performed at a prescribed percentage of 1-RM may vary greatly between individuals based on anthropometrics, type of exercise, and sex of the athlete [17].

Furthermore, research investigating sex-related differences is often performed on untrained or moderately trained individuals [24]. It is therefore of interest to investigate if the association of anthropometric variables for maximal strength, relative strength, and strength endurance varies between males and females in a strength-trained population. Another gap in the literature is that sex-specific differences, especially for muscular endurance work, are commonly assessed in tests that lack ecological validity for what is being performed in training. As such, this study investigates the association of different anthropometric variables with measures of strength (maximal and relative strength and strength endurance) in the squat and bench press in strength-trained individuals matched in age and training experience and whether the association differs between the sexes. The study is valuable in comparing the association of anthropometric variables and strength performance between the sexes, as a conclusion based on a large mixed sample can lead to false conclusions since the results might be skewed due to anthropometric differences between the sexes (body height, body mass, and fat percentage). Females were hypothesized to have lower maximal and relative strength but greater strength endurance based on earlier research [8,17]. Within the sexes, lean mass and fat percentage were hypothesized to be associated with maximal and relative strength, with limb lengths inversely associated with strength endurance [13].

2. Materials and Methods

A randomized cross-over trial was performed to investigate the association between different anthropometric variables and strength in the squat and bench press for strength-trained male and female participants. To familiarize the participants with the testing procedure and establish levels of strength, all participants took part in a familiarization session > 72 h before the day of testing, consisting of an identical testing protocol. Participants were instructed not to train <24 h before testing and not to consume caffeine on the day of testing. They had to record a 24-h food log before the familiarization and were asked to replicate it to minimize variation in energy intake and hydration. The day of testing started with the participants' height and segment lengths (upper arm, lower arm, thigh, and shank) being manually measured three times to the nearest 0.1 cm by a researcher, and at least two of the measurements had to be identical for the measurement to be valid. No measure violated this requirement, with the measurements being based upon hallmarks (upper arm: acromion to the lateral epicondyle of the distal part of the humerus; lower arm: lateral epicondyle of the distal part of the humerus to the lateral epicondyle of the distal part of the ulna; upper leg: greater trochanter of the femur to the distal lateral condyle of the femur; shank: distal lateral condyle of the femur to the lateral malleolus). Afterward, participants were weighed, with body composition estimated by a calibrated Tanita bioelectrical impedance device (MC-780MA). Then, the warm-up was initiated, which was performed in a similar manner for both the back squat and bench press. Stance and grip width were measured for the squat and bench press on the day of familiarization, which were required to be similar in the 1-RM and AMRAP tests. To avoid reductions in ecological validity, the participants were not constrained in the use of equipment such as chalk, belts, lifting shoes, or wrist wraps, as long as the equipment was kept similar through all trials.

2.1. Participants

A total of 36 resistance-trained males (n = 19) and females (n = 17) with no injury or illness negatively affecting performance in the squat and bench press participated in the study. The participants were required to be >18 years old with a minimum of 12 months of consistent resistance training with >2 sessions per week. Furthermore, participants had to be able to lift 1 and 1.2 × body mass in the bench press and squat for males and 0.7 and 1 × body mass in the bench press and squat for females. The study procedure was explained both orally and in writing, and written consent had to be signed before participation. This study was approved by the local ethics committee and the Norwegian Center for Research Data, and it conformed to the latest revision of the Helsinki Declaration (project No. 445723).

2.2. Testing

The warm-up started with the participants performing a self-selected number of repetitions with a 20-kg barbell (ata Powerbar stainless steel 29 mm, ata Group AS, Asker, Norway), followed by a standardized number of repetitions at different percentages of the estimated 1-RM (8 repetitions at 40%, 6 repetitions at 60%, 3 repetitions at 70%, and 2 repetitions at 80%) [25]. The participants subsequently performed 1-RM attempts, with load increments of 0.25 to 5 kg for every successful attempt after 4 min of rest, until true 1-RM was established. Load increments were conducted with calibrated (±10 g) plates, ranging from 0.25 kg to 50 kg (ata Powerlifting Steel Plate, ata Group AS, Asker, Norway). To complete a successful lift in the squat, the participant had to descend until the trochanter major was below the patella before initiating the ascending phase. In the bench press, the barbell had to descend until it touched the chest without bouncing before ascending until the elbows were fully extended. The feet, glutes, and upper back had to be in contact with the surface and the bench throughout the lift. The technical requirements were visually controlled by an experienced strength-and-conditioning professional, while two spotters secured safety in each lift.

The 1-RM on the day of testing in the squat and bench press was used to establish the load for the AMRAP test (60% of the 1-RM), which was performed after the 1-RM test with similar technical requirements. No rest was allowed in the AMRAP test, whereby too long a pause (>1 s in the top position of the lift) resulted firstly in a warning, while a second pause of >1 s was defined as failure. The participants performed as many repetitions as possible until they were unable to complete a full repetition without assistance from the spotters.

2.3. Statistical Analysis

Descriptive statistics are presented as means and standard deviations. The intraclass correlation coefficient (ICC) from the familiarization day to the test day was calculated to investigate the reliability in the squat and bench press when performing the 1-RM and AMRAP tests, in which the interpretation of the ICC was that values between 0.5 and 0.75 indicated moderate reliability, between 0.75 and 0.9 good reliability, and above 0.9 excellent reliability [26]. Between-group differences were tested by the independent samples t-test. The assumption of normality was assessed with the Shapiro–Wilk test. When the assumption of normality was violated, the non-parametric Mann–Whitney U test was used. Between-group effects were calculated according to Cohen's d ($\frac{M1-M2}{Pooled\ STD}$). Effect sizes were defined as follows: 0.01 to 0.2 = very small; 0.2 to 0.5 = small; 0.5 to 0.8 = moderate; >0.8 = large; >1.2 = very large; and >2 = huge [27,28]. The correlation between performance and anthropometric variables was calculated with Pearson's correlation coefficient. When the assumption of normality was violated, Spearman's rho was used. The strength of association was defined by the following r value: 0.1 to 0.3 = small; 0.3 to 0.5 = moderate; 0.5 to 0.7 = large; and 0.7 $\geq$ very large. The Holm–Bonferroni correction was assessed to reduce the type I error rate for the number of correlational tests performed. The between-group difference in correlation coefficients was calculated by Fisher´s Z-test with an online calculator [29]. Relative strength in the squat and bench press was calculated as external load lifted/body mass (kg). All tests were performed in SPSS v.27 (IBM Corp., Armonk, NY, USA). The level of significance was set at $p < 0.05$.

3. Results

The ICC from the familiarization day to the test day revealed good-to-excellent reliability in the squat and bench press when performing the 1-RM and AMRAP tests (ICC $\geq$ 0.76). Significant differences were observed between males and females for all measures of anthropometrics and strength performance (d $\geq$ 0.87, $p \leq 0.05$), except for age, training experience, and the number of repetitions in the AMRAP squat (d $\leq$ 0.68, $p \geq 0.07$) (Table 1). Males were taller and heaver and had longer upper and lower limbs, a lower fat percentage, more lean mass, and higher absolute 1-RM performance in the squat and bench press than females. However, females had significantly more repetitions in the AMRAP bench press test than males (Table 1).

Table 1. Descriptive statistics of the male and female participants.

	Males (n = 19)	Females (n = 17)	Difference (%)	Effect Size (d)
Age (years)	24.3 ±3.5	22.1 ± 3	9.3	0.70
Height (cm)	182 ± 7.3	166.1 ± 3.7	8.7 *	2.87
Body mass (kg)	87.1 ± 13.3	65.5 ± 5.6	24.8 *	2.29
Training years (n of years)	4.2 ± 2.5	4.7 ± 2.3	10.8	0.21
Lean mass (kg)	67.4 ± 6.1	46.3 ± 3.6	31.3 *	4.35
Fat percentage (%)	17.3 ± 6	25.1 ± 6	31.1 *	1.30
Upper arm length (cm)	33.5 ± 3.2	30.7 ± 30.4	8.6 *	0.87
Lower arm length (cm)	28.9 ± 6.7	25.1 ± 1.6	13.4 *	0.93
Thigh length (cm)	42.4 ± 2.4	38.9 ± 3.2	8.3 *	1.28
Shank length (cm)	43.9 ± 2.6	39.7 ± 1.7	9.6 *	1.97
1-RM bench press (kg)	110.7 ± 24.3	54.6 ± 10	50.7 *	3.27
AMRAP bench press (n)	17.7 ± 2.6	20.9 ± 3.5	15.6 *	1.06
Relative strength bench press	1.3 ± 0.2	0.8 ± 0.2	34.8 *	2.25
1-RM squat (kg)	146 ± 34.9	88.6 ± 17.3	39.3 *	2.20
AMRAP squat (n)	18.4 ± 3.2	21.4 ± 5.8	14.2	0.68
Relative strength squat	1.7 ± 0.3	1.4 ± 0.3	19.2 *	1.14

* indicates a significant difference between males and females at a $p < 0.05$ level.

For all participants, lean mass revealed the greatest association with 1-RM performance in both the squat and bench press ($r \geq 0.81$, $p \leq 0.01$). Furthermore, body height was associated with increased 1-RM in the squat and bench press and relative strength in the bench press ($r \geq 0.55$, $p \leq 0.01$), but it was inversely associated with AMRAP performance in both the squat and bench press ($r \leq -0.36$, $p \leq 0.02$) (Table 2).

Table 2. Correlations between different performances in the squat and bench press with anthropometric data for all male and female participants.

	All Participants						
	Bench Press				**Barbell Back Squat**		
	1-RM	**AMRAP**	**Relative Strength**		**1-RM**	**AMRAP**	**Relative Strength**
Lean mass	0.86 *	−0.25	0.72 *	Lean mass	0.81 *	−0.23	0.42
Fat percentage	−0.49 *	0.34	0.52 *	Fat percentage	−0.25	−0.10	−0.41 *
Body height	0.75 *	−0.36	0.55 *	Body height	0.66 *	−0.41 *	0.24
Grip width/height	0.06	0.39	0.52	Stance width/height	0.15	−0.41 *	−0.03
Upper arm length	0.21	−0.02	0.06	Thigh length	0.46 *	−0.18	0.19
Lower arm length	0.40 *	0.01	0.11	Shank length	0.64 *	−0.26	0.43 *
	Males						
	1-RM	**AMRAP**	**Relative Strength**		**1-RM**	**AMRAP**	**Relative Strength**
Lean mass	0.29	0.03	−0.21 †	Lean mass	0.46	−0.27 †	−0.01
Fat percentage	0.82 *†	0.07	0.49 †	Fat percentage	0.61 *†	0.08	0.22 †
Body height	−0.16	0.06	−0.59 *	Body height	0.09	−0.50	−0.42
Grip width/height	0.04	0.14	0.07	Stance width/height	−0.01	0.21	0.09
Upper arm length	−0.42	−0.18	−0.41	Thigh length	0.33	−0.67 *†	0.07
Lower arm length	−0.04	0.28	−0.22	Shank length	0.23	−0.48	−0.16
	Females						
	1-RM	**AMRAP**	**Relative Strength**		**1-RM**	**AMRAP**	**Relative Strength**
Lean mass	0.75 *	0.49	0.52 †	Lean mass	0.57 *	0.58 †	0.39
Fat percentage	−0.47 †	0.13	−0.74 *†	Fat percentage	−0.55 †	−0.60 *	−0.67 *†
Body height	0.28	0.27	−0.08	Body height	−0.05	−0.07	−0.17
Grip width/height	0.48	0.59 *	0.43	Stance width/height	−0.47	0.02	−0.53
Upper arm length	−0.09	0.36	−0.24	Thigh length	0.03	0.06 †	0.19
Lower arm length	0.14	0.37	−0.23	Shank length	0.33	0.14	0.32

* indicates a significant correlation coefficient at a $p < 0.05$ level. † indicates a significantly different correlation coefficient between males and females at a $p < 0.05$ level.

When analyzing between the sexes, significantly different correlation coefficients were observed between males and females for fat percentage in 1-RM strength and relative strength in the bench press and squat, lean mass in relative strength for the bench press, and thigh length in the AMRAP test for the squat (Z-score ≥ -0.198, $p \leq 0.05$) (Table 2). Males increased absolute and relative 1-RM squat and bench press performance with increasing fat percentage, while women increased this performance with decreasing fat percentage (Figure 1).

Furthermore, females showed a positive correlation between increased lean body mass and AMRAP squat and relative 1-RM bench press performances, while males showed a negative correlation with these two parameters. In addition, the number of repetitions in the AMRAP squat test increased when thigh length was shorter in males, while no correlation was found in females (Figure 2).

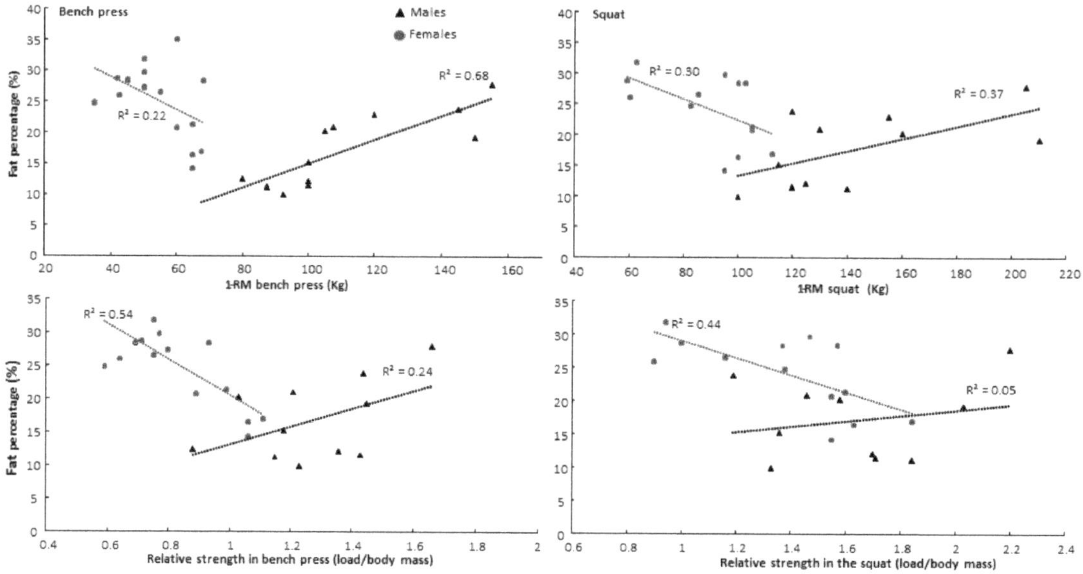

Figure 1. Sex-specific association between fat percentage with measures of maximal and relative strength in squat and bench press.

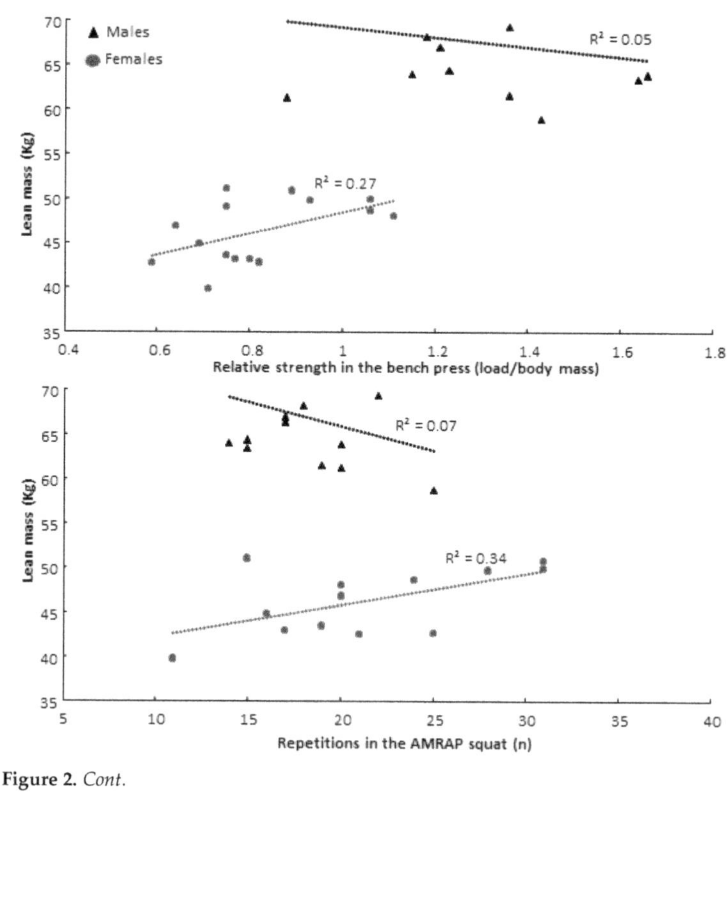

Figure 2. *Cont.*

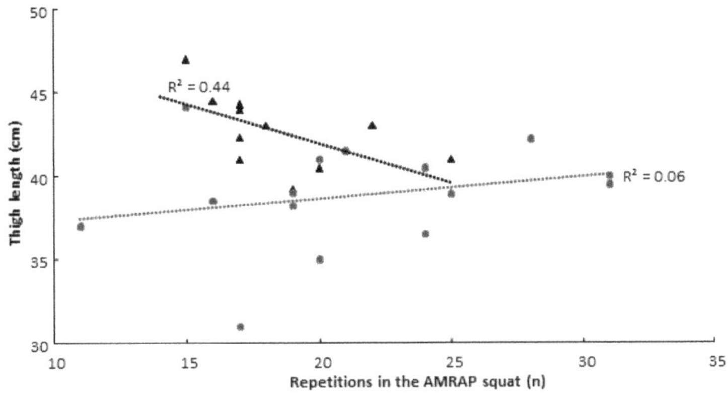

Figure 2. Association of lean mass and thigh length with relative strength in the bench press and the number of repetitions performed in the AMRAP squat for male and female participants.

4. Discussion

The current study aimed to compare the associations of anthropometric variables with measures of strength between resistance-trained males and females, matched in chronological age and training experience. It was hypothesized that females possessed lower maximal and relative strength and greater strength endurance, and limb lengths were hypothesized to be inversely associated with strength endurance within the sexes. The hypotheses were only partially confirmed. Unsurprisingly, males were significantly stronger in the squat and bench press in both maximal and relative strength. The differences in stature and body composition between the sexes most likely explain the strong association in all participants for height and lean mass with 1-RM strength. Earlier research suggests that heavier individuals with more muscle mass are generally stronger [30], and the males in this study were taller and heavier with more lean body mass in comparison to the females. Body height differences between males and females may also account for the inverse relationship between height and the AMRAP tests for all participants, as females performed significantly more repetitions in the AMRAP bench press and showed a trend towards significantly more repetitions in the AMRAP squat.

A significantly different correlation coefficient was observed for males and females between fat percentage and 1-RM performance and relative strength in both the squat and bench press. The results suggest increased fat percentage is associated with greater 1-RM strength in males but reduced relative strength in females. Increased fat percentage may contribute to the 1-RM bench press as a larger body mass could make the range of motion for the lift shorter (less work) by reducing bar path displacement from lockout to the bottom position of the lift. In the squat, the extra mass may aid in stabilizing the bar [30]. Unexpectedly, fat percentage in females trended towards being inversely associated with 1-RM strength, contradicting the association observed in males and earlier research [7,30]. Fat percentage in females was furthermore observed to be inversely associated with relative strength for both the squat and the bench press, as increasing body mass is suggested to increase the absolute weight lifted while decreasing relative strength [31], especially if mass increases are composed of non-contractile tissue.

Although greater lean mass is known to be associated with increased strength in powerlifting [32], a significantly different correlation coefficient was observed for males and females between lean mass and relative strength in the bench press, indicating lean mass to be a greater predictor of relative bench press strength in females compared to males. The finding was unexpected, as force per unit of muscle mass has been suggested to be similar between males and females [33]. Why this difference is observed cannot be stated for certain, although force per unit of muscle mass could be lower with increased muscle mass [33] and the difference in muscle mass between the sexes is commonly greater in the upper body [21]. Furthermore, lean mass was only significantly associated with increased 1-RM strength for all participants and females in the current study (Table 2). The trivial correlation in males may be due to sample size and variance in anthropometric variables, although variations in experience with the squat and bench press cannot be discounted as a confounding variable as neurological adaptations may affect the association between lean mass and 1-RM strength [33].

The different associations in strength performance with anthropometric variables observed between sexes are possibly a result of different training motives [34,35]. It could be that the female participants with lower fat percentages are also most devoted to resistance training, as weight management is one of several motivational factors for habitual resistance training in females [36,37]. In males, on the other hand, "being strong" is a traditionally valued trait [34], which at a certain point may come at the cost of increased body mass to further increase maximal strength [31]. Furthermore, females have been suggested to be more aware of self-presentation [35], whereby a certain "ideal" body composition might be a result of sex expectations [34]. The assumption of different motives is further supported by fat percentage being inversely associated with AMRAP squat performance in females (Table 2), as earlier research has associated repeated squat performance with VO2max while being inversely associated with body mass and fat percentage [38]. Therefore, the different associations between the sexes might be a result of training history [33]. However, these interpretations must be evaluated with caution, as training status, loading ranges, and type of exercise are all factors that may influence the observed association [20].

A significantly different correlation coefficient was observed for males and females between thigh length and the number of repetitions in the AMRAP squat (Figure 2). The results suggest that increased thigh length is a greater predictor of a decreased number of repetitions performed in the AMRAP squat for males compared to females, although an inverse association was expected in both sexes based on earlier research [13], as long femurs will increase the work performed per repetition. The difference might be a result of the males lifting greater absolute loads in the squat along with their greater body mass (Table 1). Thus, the absolute external torque requirements will be greater in males, possibly making each sub-maximal repetition relatively more fatiguing. This was not observed in the bench press, as the lengths of the upper and lower arm were not significantly associated with bench press performance, a relationship that has been found to vary in the literature based on the population studied [30]. Lastly, grip width relative to height was significantly associated with the number of repetitions performed by females in the AMRAP bench press. Increasing grip width might therefore be beneficial for females in the AMRAP bench press, possibly as a result of reducing work per repetition [19].

This study has some limitations that must be addressed. Firstly, the study would benefit from a larger sample size, as several moderate-to-large, non-significant r values were observed. A replication with a larger sample size is warranted. Secondly, lean mass accounts for all fat-free mass, such as water content and bone density. Therefore, individuals with greater body height/segment lengths might also have greater lean mass without having more contractile tissue. Thirdly, this study would be strengthened with the measurement of explanatory variables such as work being performed, cross-sectional area, and VO2max. Such measurements are warranted in future research to provide greater certainty in cause and effect. Lastly, this study used loads of 60% of 1-RM and might not be generalizable to other loading ranges.

5. Conclusions

Although cause and effect cannot be stated, fat percentage in females and thigh length in males are anthropometric variables indicated to influence appropriate loading ranges in the squat. It could be beneficial to evaluate these considerations in resistance training protocols when utilizing the barbell back squat for a strength-specific training goal. If maximal strength in the bench press is the training goal, males might benefit from increasing body mass. Increased bench press performance for females in this study was associated with greater lean mass but not increased fat percentage. However, it is important to underline that the sex-specific differences observed in this study may be influenced by the small sample size and training histories between the sexes. As such, the findings from this study should encourage caution when using small samples of mixed sexes to determine associations between strength and anthropometric variables.

Author Contributions: Conceptualization, H.N.F., S.L. and R.v.d.T.; formal analysis, M.E.H., H.N.F., S.L. and R.v.d.T.; methodology, M.E.H. and R.v.d.T.; writing—original draft, H.N.F.; writing—review and editing, S.L., M.E.H. and R.v.d.T. All authors have read and agreed to the published version of the manuscript.

Funding: This research received no external funding.

Institutional Review Board Statement: The study was conducted according to the ethical regulations for research and approved by the Norwegian Center for Research Data (project No. 445723 on 10 December 2021).

Informed Consent Statement: Informed consent was obtained from all subjects involved in the study.

Data Availability Statement: The raw data supporting the conclusion of this article will be made available by the authors without undue reservation.

Conflicts of Interest: The authors declare no conflict of interest.

References

1. Westcott, W.L. Resistance training is medicine: Effects of strength training on health. *Curr. Sports Med. Rep.* **2012**, *11*, 209–216. [CrossRef]
2. Winett, R.A.; Carpinelli, R.N. Potential health-related benefits of resistance training. *Prev. Med.* **2001**, *33*, 503–513. [CrossRef] [PubMed]
3. Kraemer, W.J.; Ratamess, N.A.; French, D.N. Resistance training for health and performance. *Curr. Sports Med. Rep.* **2002**, *1*, 165–171. [CrossRef]
4. Falch, H.N.; Rædergård, H.G.; van den Tillaar, R. Effect of different physical training forms on change of direction ability: A systematic review and meta-analysis. *Sports Med.-Open* **2019**, *5*, 53. [CrossRef] [PubMed]
5. Falch, H.N.; Haugen, M.E.; Kristiansen, E.L.; van den Tillaar, R. Effect of Strength vs. Plyometric Training upon Change of Direction Performance in Young Female Handball Players. *Int. J. Environ. Res. Public Health* **2022**, *19*, 6946. [CrossRef]
6. Rædergård, H.G.; Falch, H.N.; Tillaar, R.V.D. Effects of Strength vs. Plyometric Training on Change of Direction Performance in Experienced Soccer Players. *Sports* **2020**, *8*, 144. [CrossRef] [PubMed]
7. Suchomel, T.J.; Nimphius, S.; Stone, M.H. The importance of muscular strength in athletic performance. *Sports Med.* **2016**, *46*, 1419–1449. [CrossRef] [PubMed]
8. Hicks, A.L.; Kent-Braun, J.; Ditor, D.S. Sex differences in human skeletal muscle fatigue. *Exerc. Sport Sci. Rev.* **2001**, *29*, 109–112. [CrossRef] [PubMed]
9. Nimphius, S.; Callaghan, S.J.; Bezodis, N.E.; Lockie, R.G. Change of direction and agility tests: Challenging our current measures of performance. *Strength Cond. J.* **2018**, *40*, 26–38. [CrossRef]
10. Watts, D. A brief review on the role of maximal strength in change of direction speed. *J. Aust. Strength Cond.* **2015**, *23*, 100–108.
11. Bourgeois, F.; McGuigan, M.; Gill, N.; Gamble, G. Physical characteristics and performance in change of direction tasks: A brief review and training considerations. *J. Strength Cond. Res.* **2017**, *25*, 104–117.
12. Tibana, R.A.; de Sousa Neto, I.V.; Sousa, N.M.F.D.; Romeiro, C.; Hanai, A.; Brandão, H.; Dominski, F.H.; Voltarelli, F.A. Local Muscle Endurance and Strength Had Strong Relationship with CrossFit® Open 2020 in Amateur Athletes. *Sports* **2021**, *9*, 98. [CrossRef]
13. Cooke, D.M.; Haischer, M.H.; Carzoli, J.P.; Bazyler, C.D.; Johnson, T.K.; Varieur, R.; Zoeller, R.F.; Whitehurst, M.; Zourdos, M.C. Body mass and femur length are inversely related to repetitions performed in the back squat in well-trained lifters. *J. Strength Cond. Res.* **2019**, *33*, 890–895. [CrossRef] [PubMed]
14. Janssen, I.; Heymsfield, S.B.; Wang, Z.; Ross, R. Skeletal muscle mass and distribution in 468 men and women aged 18–88 yr. *J. Appl. Physiol.* **2000**, *89*, 81–88. [CrossRef] [PubMed]
15. Miller, A.E.J.; MacDougall, J.; Tarnopolsky, M.; Sale, D. Gender differences in strength and muscle fiber characteristics. *Eur. J. Appl. Physiol.* **1993**, *66*, 254–262. [CrossRef]
16. Maughan, R.; Harmon, M.; Leiper, J.; Sale, D.; Delman, A. Endurance capacity of untrained males and females in isometric and dynamic muscular contractions. *Eur. J. Appl. Physiol.* **1986**, *55*, 395–400. [CrossRef]
17. Hunter, S.K. Sex differences and mechanisms of task-specific muscle fatigue. *Exerc. Sport Sci. Rev.* **2009**, *37*, 113. [CrossRef]

18. Simoneau, J.-A.; Bouchard, C. Human variation in skeletal muscle fiber-type proportion and enzyme activities. *Am. J. Physiol. Endocr. Metabol.* **1989**, *257*, E567–E572. [CrossRef]
19. Wagner, L.L.; Evans, S.A.; Weir, J.P.; Housh, T.J.; Johnson, G.O. The effect of grip width on bench press performance. *J. Appl. Biomech.* **1992**, *8*, 1–10. [CrossRef]
20. Hoeger, W.W.; Hopkins, D.R.; Barette, S.L.; Hale, D.F. Relationship between Repetitions and Selected Percentages of One Repetition Maximum: A Comparison between. *J. Appl. Sport Sci. Res.* **1990**, *4*, 48.
21. Haff, G.G.; Triplett, N.T. *Essentials of Strength Training and Conditioning*, 4th ed.; Human Kinet: Champaign, IL, USA, 2015.
22. Bishop, P.; Cureton, K.; Collins, M. Sex difference in muscular strength in equally-trained men and women. *Ergonomics* **1987**, *30*, 675–687. [CrossRef] [PubMed]
23. Zourdos, M.C.; Klemp, A.; Dolan, C.; Quiles, J.M.; Schau, K.A.; Jo, E.; Helms, E.; Esgro, B.; Duncan, S.; Merino, S.G. Novel resistance training–specific rating of perceived exertion scale measuring repetitions in reserve. *J. Strength Cond. Res.* **2016**, *30*, 267–275. [CrossRef] [PubMed]
24. Roberts, B.M.; Nuckols, G.; Krieger, J.W. Sex differences in resistance training: A systematic review and meta-analysis. *J. Strength Cond. Res.* **2020**, *34*, 1448–1460. [CrossRef] [PubMed]
25. Gomo, O.; Van Den Tillaar, R. The effects of grip width on sticking region in bench press. *J. Sports Sci.* **2016**, *34*, 232–238. [CrossRef] [PubMed]
26. Koo, T.K.; Li, M.Y. A Guideline of Selecting and Reporting Intraclass Correlation Coefficients for Reliability Research. *J. Chiroprac. Med.* **2016**, *15*, 155–163. [CrossRef]
27. Cohen, J. *Statistical Power Analysis for the Behaviour Sciences (Revised Edition)*; Routledge: New York, USA, 1988; p. 74.
28. Sawilowsky, S.S. New effect size rules of thumb. *J. Mod. Appl. Stat. Methods* **2009**, *8*, 597–599. [CrossRef]
29. Soper, D.S. Significance of the Difference between Two Correlations Calculator. Available online: https://www.danielsoper.com/statcalc (accessed on 18 November 2022).
30. Ferland, P.-M.; Pollock, A.; Swope, R.; Ryan, M.; Reeder, M.; Heumann, K.; Comtois, A.S. The relationship between physical characteristics and maximal strength in men practicing the back squat, the bench press and the deadlift. *Int. J. Exerc. Sci.* **2020**, *13*, 281.
31. Latella, C.; Van den Hoek, D.; Teo, W.-P. Factors affecting powerlifting performance: An analysis of age-and weight-based determinants of relative strength. *Int. J. Perf. Anal. Sport* **2018**, *18*, 532–544. [CrossRef]
32. Brechue, W.F.; Abe, T. The role of FFM accumulation and skeletal muscle architecture in powerlifting performance. *Eur. J. Appl. Physiol.* **2002**, *86*, 327–336. [CrossRef]
33. Jones, E.J.; Bishop, P.A.; Woods, A.K.; Green, J.M. Cross-sectional area and muscular strength. *Sports Med.* **2008**, *38*, 987–994. [CrossRef]
34. Johansson, T. Gendered spaces: The gym culture and the construction of gender. *Young* **1996**, *4*, 32–47. [CrossRef]
35. Zach, S.; Adiv, T. Strength training in males and females: –Motives, training habits, knowledge, and stereotypic perceptions. *Sex Roles* **2016**, *74*, 323–334. [CrossRef]
36. Vasudevan, A.; Ford, E. Motivational Factors and Barriers Towards Initiating and Maintaining Strength Training in Women: A Systematic Review and Meta-synthesis. *Prev. Sci.* **2021**, *23*, 674–695. [CrossRef] [PubMed]
37. Twamley, E.W. *Gender, Exercise Motives, and Exercise Behavior*; Arizona State University: Tempe, AZ, USA, 2000.
38. Vaara, J.P.; Kyröläinen, H.; Niemi, J.; Ohrankämmen, O.; Häkkinen, A.; Kocay, S.; Häkkinen, K. Associations of maximal strength and muscular endurance test scores with cardiorespiratory fitness and body composition. *J. Strength Cond. Res.* **2012**, *26*, 2078–2086. [CrossRef] [PubMed]

Journal of
Functional Morphology and Kinesiology

MDPI

Article

Relationship between Maximum Force–Velocity Exertion and Swimming Performances among Four Strokes over Medium and Short Distances: The Stronger on Dry Land, the Faster in Water?

Vincenzo Sorgente [1,*], Aaron Agudo-Ortega [2], Alejandro Lopez-Hernandez [3], Jesus Santos del Cerro [2], Diego Minciacchi [1,*] and José María González Ravé [2]

1 Kinesiology and Motor Control (Ki.Mo.Co.) Laboratory, Department of Experimental and Clinical Medicine, Physiological Sciences Section, University of Florence, 50134 Florence, Italy
2 Sport Training Laboratory, Faculty of Sports Sciences, University of Castilla-La Mancha, 45071 Toledo, Spain
3 Castilla-la-Mancha Swimming Federation, Castilla-la-Mancha, 02001 Albacete, Spain
* Correspondence: vincenzo.sorgente@unifi.it (V.S.); diego.minciacchi@unifi.it (D.M.)

Abstract: Evaluating force–velocity characteristics on dry-land is of the utmost importance in swimming, because higher levels of these bio-motor abilities positively affect in-water performance. However, the wide range of possible technical specializations presents an opportunity for a more categorized approach that has yet to be seized. Therefore, the aim of this study was to identify feasible differences in maximum force–velocity exertion based on swimmers' stroke and distance specialization. To this scope, 96 young male swimmers competing at the regional level were divided into 12 groups, one for each stroke (butterfly, backstroke, breaststroke, and front crawl) and distance (50 m, 100 m, and 200 m). They performed two single pull-up tests, 5-min before and after competing in a federal swimming race. We assessed force (N) and velocity (m/s) exertion via linear encoder. There were no significant differences between pre-post maximum force–velocity exertions, despite the decreasing trend. Force-parameters highly correlated with each other and with the swimming performance time. Moreover, both force ($t = -3.60$, $p < 0.001$) and velocity ($t = -3.90$, $p < 0.001$) were significant predictors of swimming race time. Sprinters (both 50 m and 100 m) of all strokes could exert significantly higher force–velocity compared to 200 m swimmers (e.g., 0.96 ± 0.06 m/s performed by sprinters vs. 0.66 ± 0.03 m/s performed by 200 m swimmers). Moreover, breaststroke sprinters presented significantly lower force–velocity compared to sprinters specialized in the other strokes (e.g., 1047.83 ± 61.33 N performed by breaststroke sprinters vs. 1263.62 ± 161.23 N performed by butterfly sprinters). This study could provide the foundation for future research regarding the role of stroke and distance specializations in modeling swimmers' force–velocity abilities, thus influencing paramount elements for specific training and improvement towards competitions.

Keywords: sport science; sport performance; swimming performance; training prescription; strength training; exercise physiology

Citation: Sorgente, V.; Agudo-Ortega, A.; Lopez-Hernandez, A.; Santos del Cerro, J.; Minciacchi, D.; González Ravé, J.M. Relationship between Maximum Force–Velocity Exertion and Swimming Performances among Four Strokes over Medium and Short Distances: The Stronger on Dry Land, the Faster in Water?. *J. Funct. Morphol. Kinesiol.* **2023**, *8*, 20. https://doi.org/10.3390/jfmk8010020

Academic Editor: Giuseppe Musumeci

Received: 22 December 2022
Revised: 25 January 2023
Accepted: 30 January 2023
Published: 1 February 2023

1. Introduction

Swimming can be defined as a closed-skills sport, i.e., a sporting activity in which the environment is relatively highly consistent, predictable, and self-paced for performers [1]. Within this setting, the sole sport-specific practice cannot grant the swimmers a great enough stimulus to maximize their gestures [2–4]. Therefore, along with the acquisition of certain technical skills, the appropriate development of bio-motor abilities such as strength, power, and endurance are considered the key to success in competitive swimming [5,6]. In particular, strength can be described as the ability of the neuromuscular system to produce force against external resistance [7,8] and it has been repeatedly speculated to be the crucial bio-motor ability for the development of optimal swimming performance [9–13]. In this regard, some investigators advocate the paramount role of strength with a deduction: if higher levels of strength mean a higher capacity to produce force against water resistance, then consequently this will improve the swimmers' velocity and ultimately their swimming performance [7,8,14,15]. Seeking to expand upon this compelling aspect, several scholars have in fact confirmed that there is a close relationship between the capacity of producing high forces and superior swimming performances. To this scope, a reliable tool used to assess swimming-specific strength (both on dry land as well as in the water) is the evaluation of force–velocity exertion, which basically dictates the relationship between the load lifted and the speed it can be moved [16,17]. Specifically, it has been shown that high levels of maximum strength in various exercises (e.g., bench press, pull-up, back squat, horizontal rows, etc.) correlate with many technical components of a swimming race, such as trunk stability during the stroke, gliding phase, diving phase, turning phase, stroke length, stroke frequency, and stroke index [2–4,6,7,11,18–25], ultimately translating into optimized swimming velocity [19]. Nonetheless, vertical pulling gestures are the most used and effective motions for evaluating force–velocity production in swimming performances [11,18]. In particular, it has been shown that the maximum velocity and force generated during the pull-up exercise highly correlates with swimming velocity [21]. These findings well demonstrate the effectiveness that well-developed bio-motor abilities, assessed through force–velocity parameters on dry land [20], hold on swimming performance.

However,, despite this promising body of research, the extensive heterogeneity lying within competitive swimming raises several issues that are yet to be ascertained. For instance, one critical element concerns the broad technical outlook, i.e., the four "cardinal" swimming strokes (butterfly, backstroke, breaststroke, and front crawl). Although most of the literature is focused on the front-crawl stroke [26,27], it is worth noting that the butterfly, backstroke, and breaststroke "styles" are highly specific as well as biomechanically and kinematically different from each other [28]. Based on the unique characteristics of each stroke, it could be that their differences are also reflected in the force–velocity exertion of specialized swimmers. If that is the case, it would then be possible to outline the ideal bio-motor features of swimmers who concentrate their endeavors on a particular stroke.

Another issue pertains to the numerous distances that are swum in swimming races, which dictate the athletes' specialization from a bio-energetical standpoint. On this subject, suitable research has highlighted the diverse aerobic, anaerobic, and technical demands that reside among sprint and medium- and long-distance swimming races [26,29]. Nevertheless, it should be considered that medium-distance sporting activities (as a 200 m swimming race would be defined) also seem to benefit from the high capacities of producing muscular force [30–32]. Seeing these ambiguous findings, there is a need to clarify whether measures of force–velocity significantly change depending on different swimming distances, eventually drawing out the implications in terms of building superior in-water performances based on distance specialization.

As we argued in the paragraph above, when it comes to dissecting how force–velocity levels pertain to various swimming specialties during training, the current literature presents more questions than answers. Still, these uncertain aspects provide appealing opportunities for further inquiries. With these considerations in mind, in the present study, we sought to evaluate the relationship between maximum force–velocity exertion and swimming performances in male regional-level swimmers, examining and comparing both medium (200 m) as well as short (50 and 100 m) distances; all the four strokes swum in competitions; and the amount of neuromuscular fatigue generated by the competition, collecting force–velocity data shortly before and after the swimming race. The single pull-up test, performed for one repetition exerting maximum force, was used to evaluate the swimmers' force–velocity production, whereas the official time of the respective swimming race (swum in a short course of 25 m) was considered as an indicator for swimming performance.

Ultimately, the purpose of this article is to clearly define to what extent maximum force–velocity capacities and swimming performances relate to each other when considering the intrinsically different bio-motor and technical facets of athletes specialized in a certain stroke and distance, also exploring whether and how this relationship is altered due to neuromuscular fatigue after performance. Overall, this additional knowledge would make it possible to draw evidence-based indications in order to further elevate swimming performance, favoring both academics for more profound investigations as well as coaches and athletes striving to attain ever-better competitive forms.

2. Materials and Methods

2.1. Participants

A homogeneous group of 96 male swimmers competing at the regional level took part in the study (16 ± 1.3 years of age; height of 175 ± 2.7 cm; weight of 69 ± 2.2 kg; 6.5 ± 1.1 years of experience; 466 ± 21 Fédération Internationale De Natation points of best competitive performance). Specifically, we divided the subjects per stroke (i.e., butterfly, backstroke, breaststroke, and front-crawl) and distance (i.e., 50, 100, and 200 m), therefore assigning eight subjects per group. All of the participants reported no physical injuries prior to or during the duration of the study. Moreover, the subjects were already familiar with the pull-up motion from their previous training experience. The participants were all tested individually. All of the subjects provided assent and the parents/guardians provided informed consent after a detailed description of the study procedures. The study was approved by the local Ethics Committee of the university (FGM02102019) and was conducted in accordance with the Declaration of Helsinki.

2.2. Data Collection

We collected data regarding the in-water performances and the force–velocity parameters. As for the in-water performance, swimming race times were collected during federal swimming races by professional personnel employed in the local swimming federation. As for the force–velocity parameters in the ascending phase of the single pull-up test, velocity (m/s) and force (N) were collected using the Vitruve linear encoder (Speed4Lifts, Madrid, Spain). Specifically, this linear encoder comes in the portable form of an 8 cm^3 box, equipped with an extensible wire that is attachable via a Velcro strap. Moreover, the Vitruve linear encoder is embedded with a smartphone app that allows for insertion of the subject's height and weight, consequently calculating specific performances in a selected exercise (in this case, the pull-up). In particular, we used the velocity-based data registered by the encoder (i.e., power and velocity) to provide the force values.

2.3. Procedures

The subjects performed the single pull-up tests and concurrent swimming race at the end of their preparatory cycle of training (i.e., 8 weeks after the start of the season). The experiments were performed inside a regular, short-course (25 m) competitive swimming facility, during competition days. In particular, the single pull-up tests were performed in a large, quiet room inside the facility and near the pool. Furthermore, the pull-ups were performed using a standard steel bar of 3.81 cm in diameter (1.5 inches), standing 2.50 m from the ground.

Prior to the start of the experiment, the participants were advised to perform two single pull-up tests, with the first test taking place 5 min before the swimming competition and repeating the same test 5 min after the aforementioned competition. Specifically, the subjects were instructed to perform one repetition of the pull-up motion, exerting the highest force possible (i.e., pulling as strong and fast as they can). Moreover, the subjects had to follow precise criteria regarding the pull-up execution; first, they had to reach for the bar with a prone grip, without their feet touching the ground and by maintaining their arms and elbows straight. This was considered the starting position of the pull-up test. From this hanging position, after a brief verbal cue ("Ready, go"), the subject would then perform the pull-up, which had to be executed without any movement of the legs and passing with the chin over the bar. To ensure the procedure for measuring pull-ups

5 min before and 5 min after the swimming races, we employed a stopwatch. According to Kraemer and Fleck [33], 5 min is considered a long rest period, capable of dissipating the amount of fatigue experienced during anaerobic physical exertions. According to this information, we used the 5 min rest period to ensure enough recovery pre-competition and to also establish the same rest period after the competition.

In order to prepare for the competitions, the subjects first executed 20 min of warm-up. Specifically, the warm-up was 20 min long and consisted in the first part (about 5 min) being performed on dry land using body weight (e.g., squats) and elastic bands exercises (e.g., shoulders horizontal internal and external rotation), whereas the second part (about 15 min) was performed in-water and included sport-specific drills and exercises (i.e., turns, underwater glides, swimming at various paces and stroke rhythms), performed at light intensity. After 20 min of warm-up (i.e., 5 min before the respective swimming race), the subject came into the testing room. In order to register the force–velocity data, the linear encoder was attached to the subjects' hips through a harness. The linear encoder was instead attached to the ground, within the same vertical plane as the subjects. In this way, it was possible to collect accurate data with minimal invasiveness. After that, the subjects performed the single pull-up test according to the criteria explained above. Five minutes after the single pull-up test, the subjects took part in the federal swimming race assigned. Each swimmer took part in only one race. Another five minutes after the swimming race, the subjects returned and performed a second single pull-up test following the same experimental setup. Each swimmer that was recruited for this study was tested individually for both the dry land and in-water performance assessments. However, given the competitive nature of this experimental setting, the subjects competed in the swimming races with other athletes who did not take part in this study.

2.4. Statistics

The data were analyzed using IBM SPSS Statistics 26 software. Parametric analyses were conducted as the Shapiro–Wilk test revealed a normal distribution of data ($p > 0.05$). First, we checked for test–retest reliability taking advantage of the two force–velocity assessments we made within this experimental setup. The resulting correlation coefficient was 0.81, therefore indicating an instance of good test–retest reliability. Then, detriments in the pull-up performance due to neuromuscular fatigue were sought using the ANOVA repeated measures test for both the velocity and the force generated before and after the swimming race. Furthermore, the Pearson correlation coefficient was used to define the levels of dependence among the force and the velocity in the single pull-up test and the time of the swimming races. Moreover, we used a multiple linear regression model to quantify the relationship between the velocity and force in the single pull-up test with the time of the swimming races. Finally, we implemented one-way ANOVA followed by the Bonferroni post-hoc test for multiple comparisons in order to detect differences in velocity and/or force exerted in the single pull-up test among strokes and distance specialties.

3. Results

The fatigue generated by the swimming performances affected both the velocity as well as the force in the single pull-up test (Table 1). Specifically, the pre-post percentage difference in velocity ranged from −1.71% in the 100 m backstroke group to −2.86% in the 50 m breaststroke group, whereas the percentage difference in force ranged from −2.43% in the 100 m breaststroke group to −3.39% in the 100 m backstroke group. However, the ANOVA repeated measures test reported no statistically significant differences in either velocity ($F(96, 1) = 1.89$, $p = 0.17$) and force ($F(96, 1) = 2.33$, $p = 0.35$) among the groups. While including a control group to test for fatigue after the swimming race would improve the experimental design, we have no reason to believe that the general loss in force–velocity post-competition was not due to the swimming race, which was the only physical stimulus occurring between the pre- and post-evaluations. Moreover, we employed more than enough recovery time regarding the single pull-up tests before and after the swimming race (i.e., 5 min) [33]. Consequently, it was reasonable to expect similar values of force–velocity performances from the subjects, which was only partially the case. However, we again specify that this trend did not achieve statistical significance, and thus can only be seen as a speculative interpretation of the phenomenon.

Table 1. Descriptive table of results.

		Velocity (m/s)			Force (N)		
Group	T (s)	Pull-Up Pre	Pull-Up Post	Fatigue (%)	Pull-Up Pre	Pull-Up Post	Fatigue (%)
Bu 50	27.35 ± 1.14	0.97 ± 0.07	0.95 ± 0.07	−2.33	1263.62 ± 161.23	1223.97 ± 155.62	−3.09
Ba 50	29.88 ± 0.96	0.95 ± 0.06	0.93 ± 0.06	−2.5	1230.66 ± 116.82	1194.58 ± 111.24	−2.88
Br 50	32.19 ± 0.89	0.79 ± 0.03	0.76 ± 0.03	−2.86	1047.83 ± 61.33	1016.11 ± 66.81	−2.95
FC 50	24.86 ± 0.87	0.96 ± 0.06	0.94 ± 0.05	−2.15	1247.48 ± 137.14	1211.46 ± 140.39	−2.85
Bu 100	60.49 ± 1.16	0.88 ± 0.02	0.86 ± 0.02	−2.27	1175.36 ± 53.48	1146.21 ± 55.09	−2.45
Ba 100	63.82 ± 2.32	0.88 ± 0.05	0.86 ± 0.05	−1.71	1151.48 ± 31.28	1111.98 ± 29.45	−3.39
Br 100	71.72 ± 2.19	0.77 ± 0.03	0.75 ± 0.03	−2.25	1024.42 ± 29.51	999.12 ± 26.57	−2.43
FC 100	53.63 ± 1.18	0.93 ± 0.04	0.92 ± 0.03	−1.72	1188.69 ± 55.30	1156.16 ± 58.93	−2.68
Bu 200	151.16 ± 13.33	0.66 ± 0.03	0.65 ± 0.04	−2.28	963.63 ± 28.62	925.44 ± 31.56	−3.9
Ba 200	144.79 ± 8.13	0.67 ± 0.03	0.65 ± 0.03	−2.61	960.82 ± 29.90	930.14 ± 28.98	−3.11
Br 200	158.45 ± 4.49	0.67 ± 0.03	0.66 ± 0.03	−2.24	957.55 ± 34.13	932.50 ± 36.60	−2.56
FC 200	120.63 ± 4.71	0.69 ± 0.04	0.67 ± 0.04	−2.52	966.85 ± 33.89	940.01 ± 27.95	−2.69

Abbreviations: T = time of the swimming race; Bu = butterfly; Ba = backstroke; Br = breaststroke; FC = front crawl; fatigue (%) = percentage change between pull-up pre and pull-up post the swimming race. The numbers "50", "100", and "200" next to each group indicate the distances in which the swimmers specialized.

Considering the whole sample of subjects ($n = 96$), the application of the Pearson r coefficient revealed a strong correlation between velocity and force (0.94 and 0.93 for the pull-up pre and post competition, respectively), suggesting that these two parameters may describe the same trend in this context. Likewise, the correlation between the swimming race time and force in the pull-up test was −0.74 both pre and post competition, whereas the correlation between the swimming race time and velocity during the pull-up test was −0.86 both pre and post competition, indicating that stronger/faster performances in the single pull-up test correlate with lower (thus better) swimming race times. Moreover, this strict correlation between force and velocity indicates that there is an almost linear relationship between them (i.e., more force generated means reaching a higher velocity and vice versa) (Figure 1).

Multiple linear regression was calculated to predict swimming race times based on the velocity and force generated in the single pull-up test, before the competition. In order to include all of the results in the same explanatory model, we standardized the swimming race times for each distance and stroke, calculating the respective z-scores. In a similar way, given the differences occurring in both force and velocity requirements in the single pull-up test among the experimental groups, the independent variables (i.e., velocity and force) were also standardized by z-scores. It is necessary to standardize the values since the explanatory variables in regression models have different scales and different levels of size. Considering the multiple linear regression analysis, a significant regression equation was found ($F(2, 93) = 78.17$, $p < 0.001$), with an R2 of 0.81 and an R2 adjusted of 0.80 (Table 2).

In particular, the swimmers' predicted race time was equal to 3.03×10^{-14}—0.046 (velocity)—0.037 (force). It should be noted that positive z-score values indicate values above the group mean; therefore, an increased z-score represents an increase in either the velocity or the force. Specifically, the model showed that a unitary increase in the velocity z-scores resulted in a decrease in the target variable (z-point, i.e., the swimming race time) of 0.046 s, whereas a unitary increase in the force z-scores predicted a slightly smaller decrement of 0.037 s. Bearing in mind that positive z-point values correspond to swimming race times above the mean and vice versa, the above behavior indicates that the higher the velocity or force generated by the athletes, the shorter their swimming race time. In addition, both the velocity ($t = -3.90$, $p < 0.001$) and force in the pull-up tests ($t = -3.60$, $p < 0.001$) were significant predictors of swimming race time.

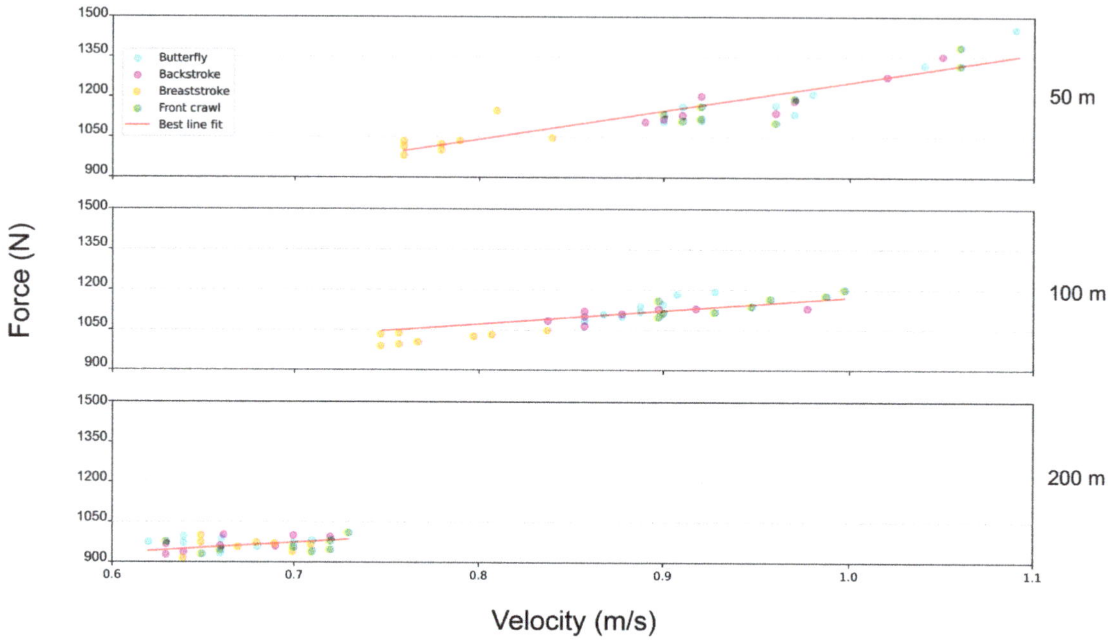

Figure 1. Scatter plot showing the maximum force–velocity exertion of the swimmers, grouped by stroke, and differentiated by swimming race distance. As shown by the best line fit, the high correlation between the force and velocity values in the single pull-up test reflects an almost linear trend. This is particularly evident within the 50 m and 100 m groups of swimmers.

Table 2. Multiple linear regression analysis to predict swimming performances based on z-scored velocity and force generated in the single pull-up test. The analysis considers both the distance and stroke groups.

ANOVA					
Model	**DF**	**Sum of Square**	**Mean Square**	**F Statistic**	***p*-Value**
Regression	2	33.77	16.88	78.17	<0.001
Residual	93	50.09	0.54		
Total	95	83.86	0.88		
Coefficients (R-square = 0.81; adjusted R-squared = 0.80)					
Model	**Estimate**	**Standard Error**	**t-Value**	***p*-Value**	
(Constant)	3.03×10^{-14}	0.075	1.08	0.28	
Velocity (z-score)	−0.046	0.126	−3.90	<0.001	
Force (z-score)	−0.037	0.121	−3.60	<0.001	

Finally, the one-way ANOVA was performed to compare the differences in velocity and force based on stroke and distance (Table 3). Regarding the validation of the test, although the a priori power was low (0.28), the null hypothesis was still rejected. Moreover, the one-way ANOVA revealed that there was a statistically significant difference in both velocity (F (11, 84) = [66.80], $p > 0.001$) as well as force among the groups of swimmers (F (11, 84) = [19.60], $p > 0.001$).

Table 3. ANOVA tables for velocity (m/s) and force (N) in the single pull-up test before the swimming competition.

Velocity (m/s)					
Source	**DF**	**Sum of Square**	**Mean Square**	**F Statistic**	***p*-Value**
Groups (between groups)	11	1.36	0.12	66.80	<0.001
Error (within groups)	84	0.16	0.0019		
Total	95	1.52	0.016		
Force (N)					
Source	**DF**	**Sum of Square**	**Mean Square**	**F Statistic**	***p*-Value**
Groups (between groups)	11	1,329,427.64	120,857.06	19.60	<0.001
Error (within groups)	84	517,981.89	6166.45		
Total	95	1,847,409.53	19,446.42		

Specifically, Table 4 reports the Bonferroni post-hoc test for multiple comparisons (Table 4). A fascinating aspect emerging from the post-hoc analysis is that almost all the significant differences between the groups in velocity corresponded to the same significant differences between groups in force, with the only exceptions being between the 200 m backstroke and the 50 m breaststroke (i.e., the mean difference in velocity was 0.12 ± 0.09 and statistically significant, whereas the mean difference in force was 87.01 ± 61.48 and not statistically significant) and between the 50 m breaststroke and 100 m front crawl (i.e., the mean difference in velocity was 0.15 ± 0.10 and not statistically significant, whereas the mean difference in force was 140.86 ± 99.54 and statistically significant). This further confirms the assumption that within this experimental setting, there is an almost linear relationship between maximum force and velocity productions.

Table 4. Bonferroni post-hoc comparisons test for velocity (m/s) and force (N) across the experimental groups. The mean differences are shown. The asterisk * shows that the mean difference is significant at the 0.05 level. Interestingly, most of the significant post-hoc differences in velocity corresponded to the same significant post-hoc differences in force among the experimental groups, further strengthening the suggestion that the velocity and force generated executing a vertical pulling motion are (almost) linearly intertwined.

Velocity (m/s)											
Group	**Ba 50**	**Br 50**	**FC 50**	**Bu 100**	**Ba 100**	**Br 100**	**FC 100**	**Bu 200**	**Ba 200**	**Br 200**	**FC 200**
Bu 50	0.02	0.20 *	0.001	0.10	0.10	0.20 *	0.04	0.31 *	0.30 *	0.30 *	0.28 *
Ba 50	0	0.17 *	0.01	0.08	0.08	0.18 *	0.02	0.29 *	0.29 *	0.28 *	0.26 *
Br 50	0.17 *	0	0.18 *	0.11 *	0.10 *	0.02	0.15 *	0.12 *	0.18 *	0.11 *	0.11 *
FC 50	0.01	0.18 *	0	0.09	0.09	0.19 *	0.03	0.30 *	0.30 *	0.29 *	0.27 *
Bu 100	0.08	0.10	0.09	0	0.004	0.11 *	0.05	0.22 *	0.21 *	0.21 *	0.19 *
Ba 100	0.08	0.09	0.08	0.003	0	0.11 *	0.05	0.21 *	0.21 *	0.20 *	0.19 *
Br 100	0.18 *	0.01	0.19 *	0.11 *	0.11 *	0	0.16 *	0.11 *	0.11 *	0.12 *	0.12 *
FC 100	0.02	0.15	0.03	0.05	0.06	0.16 *	0	0.27 *	0.27 *	0.26 *	0.24 *
Bu 200	0.29 *	0.10 *	0.30 *	0.22 *	0.21 *	0.11 *	0.27 *	0	0.003	0.08	0.03
Ba 200	0.29 *	0.10 *	0.30 *	0.21 *	0.21 *	0.10 *	0.27 *	0.003	0	0.07	0.03
Br 200	0.28 *	0.09 *	0.29 *	0.21 *	0.20 *	0.10 *	0.26 *	0.01	0.01	0	0.02
Force (N)											
Group	**Ba 50**	**Br 50**	**FC 50**	**Bu 100**	**Ba 100**	**Br 100**	**FC 100**	**Bu 200**	**Ba 200**	**Br 200**	**FC 200**
Bu 50	32.96	215.79 *	16.14	88.26	112.13	239.21 *	74.93	299.99 *	302.80 *	306.07 *	296.77 *
Ba 50	0	182.83 *	16.82	55.30	79.18	206.25 *	41.97	267.04 *	269.84 *	273.11 *	263.81 *
Br 50	182.83 *	0	199.65 *	127.54 *	103.66 *	23.42	140.86 *	84.20 *	87.01	90.28 *	80.98 *
FC 50	16.82	199.65 *	0	72.12	96	223.07 *	58.79	283.86 *	286.66 *	289.93 *	280.63 *
Bu 100	55.30	127.54	72.12	0	23.88	150.95 *	13.33	211.74 *	214.54 *	217.82 *	208.63 *
Ba 100	79.18	103.66	96	23.88	0	127.07 *	37.20	187.86 *	190.67 *	193.94 *	184.64 *
Br 100	206.25 *	23.42	223.07 *	150.95 *	127.07 *	0	164.28 *	60.79 *	63.59 *	66.86 *	57.56 *
FC 100	41.97	140.86 *	58.79	13.33	37.20	164.28 *	0	225.06 *	227.87 *	231.14 *	221.84 *
Bu 200	267.04 *	84.20 *	283.86 *	211.74 *	187.86 *	60.79 *	225.06 *	0	2.81	6.08	3.22
Ba 200	269.84 *	87.01 *	286.66 *	214.54 *	190.67 *	63.59 *	227.87 *	2.81	0	3.27	6.03
Br 200	273.11 *	90.28 *	289.93 *	217.82 *	193.94*	66.86 *	231.14 *	6.08	3.27	0	9.3

Abbreviations: Bu = butterfly; Ba = backstroke; Br = breaststroke; FC = front crawl. The asterisk * shows that the mean difference is significant at the 0.05 level.

When grouping the swimmers by stroke specialization, the 50–100 m sprinters were significantly faster and stronger in the single pull-up test than the 200 m middle-distance swimmers (Figure 2). Despite still being statistically significant, this tendency was quantitatively less prominent for the breaststroke performers. Instead, when categorizing the swimmers by the same race distance, we observed significant differences in maximum force–velocity exertion among breaststroke sprinters (i.e., 50 m and 100 m) and the other three strokes (i.e., butterfly, backstroke, and front crawl, both in 50 m as well as 100 m). Specifically, the 50 m and 100 m breaststroke performers presented significantly worse force–velocity parameters in the single pull-up test than their butterfly, backstroke, and front crawl peers (Figure 2). Regarding the 100 m swimmers, although there were no significant differences in force–velocity values compared to their 50 m counterparts, the force–velocity peaks were always reached by the 50 m sprinters. Furthermore, this behavior was not present in the middle-distance (i.e., 200 m) swimmers, which showed no significant differences from one stroke to another, although their maximum force–velocity exertions were all significantly lower compared to the 50 m and 100 m strokes.

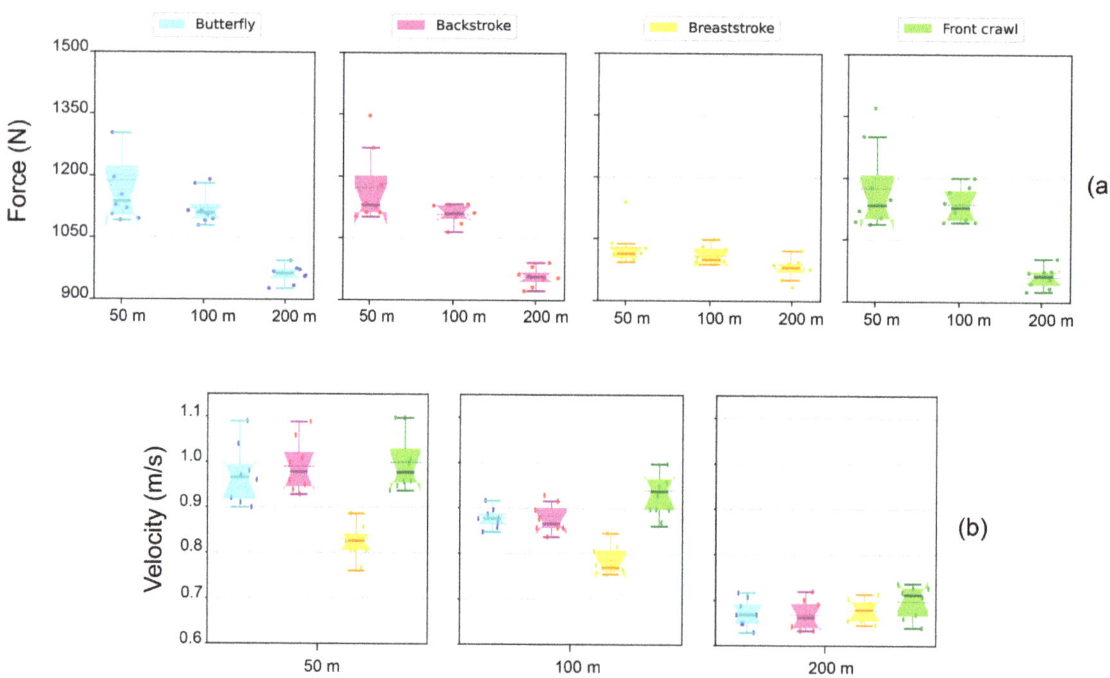

Figure 2. Scatter box plot showing the maximum force (N; panel (**a**)) and velocity (m/s; panel (**b**)) exerted in the single pull-up test, before the swimming competition. (**a**) Within all four strokes, a trend emerged where the sprinters (both 50 m and 100 m) could exert significantly higher forces compared to the medium-distance swimmers. Although it remained statistically significant, this trend was less evident regarding the breaststroke swimmers. (**b**) The 50 m and 100 m breaststroke swimmers presented significantly lower levels of velocity in the single pull-up test compared to the swimmers of other strokes competing in the same distance. However, this was not the case with the group of 200 m swimmers, where there were no significant differences among the groups.

4. Discussion

In the present study, we sought to explore the relationship between maximum force–velocity exertion and competitive performances in regional-level swimmers, specialized both for stroke (i.e., butterfly, backstroke, breaststroke, or front crawl) and race distance (i.e., 50, 100, or 200 m). The purpose of this investigation was to leverage a reliable and robust assessment method of the athletes' bio-motor abilities in order to acknowledge trends, patterns, and differences capable of bringing valuable insights for highly specialized performance enhancements in swimming.

In previous review articles, it has been argued that more accomplished swimmers presented significantly lower energy expenditure, especially among swimmers specialized in middle-distance and long-distance races [15,28,34,35]. This was due to energy expenditure being a more limiting factor in swimming races from 400 m to 1500 m than neuromuscular-fatigue-related variables in 50 m, 100 m, and 200 m swimming races. According to the work of Pyne and Sharp [34], this implied that there could be a considerable effect of stroke and distance specialization on "swimming economy" and energy system management, which could be investigated by measuring neuromuscular energy expenditure in sprinters and middle-distance swimmers. Nevertheless, although our findings highlighted a trend of a general reduction in both velocity and force generated in the single pull-up test shortly after the swimming competition (Table 1), the ANOVA repeated measures revealed no significant differences between the pre- and post evaluations (F (96, 1) = 1.89, $p = 0.17$ concerning velocity and F (96, 1) = 2.33, $p = 0.35$ concerning force, respectively). Thus, in contrast with the above-mentioned measurement (i.e., VO2 max consumption), neuromuscular energy expenditure does not seem to be a specific enough method to assert either specific performance features or differences among stroke/distance specializations in swimming.

We found a strong correlation in the form of the Pearson r coefficient between force (N) and velocity (m/s) parameters in the ascending phase of the single pull-up test (i.e., 0.94 and 0.93 for the pull-ups pre and post competition, respectively). This outcome confirms the results of other recent investigations which have suggested concentrating dry-land training efforts on enhancing the neuromuscular abilities of swimmers, particularly on the integration and coordination of musculature to perform specific tasks under high loads or in an explosive fashion [16,22,28]. Similarly, the same analysis showed a high degree of correlation between the swimmers' maximum force–velocity exertion and their respective race times (i.e., −0.74 between the force and swimming race time and −0.86 between the velocity and swimming race time). In this regard, we are in line with the research work conducted by Perez-Olea et al. [21], which showed that the 50 m front crawl swimming time was highly correlated with force–velocity variables of the ascending phase of the single pull-up test. Moreover, our multiple linear regression analysis further substantiates the validity of the pull-up motion mechanics to predict swimming performance in trained swimmers (Table 2). Hereof, the beta coefficients (i.e., velocity and force z-scores) were both negative. This means that the higher the value of these beta coefficients, the shorter the time in the swimming race, ultimately resulting in a better competitive outcome. These findings further promote the analysis of pull-up mechanics as a valid, efficient, and reliable means to both calibrate and predict crucial aspects of competitive swimming performances. Concerning this aspect, it is worth noting that we designated the swimming race times as a measure of in-water performance. However, it would be interesting to expand upon this research topic also considering more specific aspects that effectively contribute to the final performance. For instance, analyzing measures of technical proficiency such as stroke length, stroke index, stroke frequency, drag area during stroke, etc., could provide further support in understanding how maximum force–velocity exertion reshapes based on the swimmers' stroke-distance specialization and how scholars and coaches could leverage these distinctions to enhance highly specific elements of swimming performance.

Among all the strokes, the 50 m and 100 m sprinters had significantly higher force–velocity values in the single pull-up test than the 200 m middle-distance swimmers (Table 4). Nonetheless, despite maintaining statistical significance, this trend was flattened for the breaststroke performers compared to the other three strokes (Figure 2).

In terms of swimming performance optimization, we confirm that the ability to produce higher amounts of force–velocity can indeed be useful in improving swimming race times, especially in sprinters [18–21,23,28,35]. However, our results also indicated that force–velocity values tended to be lower in competitors specialized in middle-distance races (Figure 2). In this regard, it is well established that the specific contributions of various energetic systems depend on both the length of the race and the intensity of the pace used [33]. Specifically, middle-distance competitors may prioritize the maximization of aerobic capacities in lieu of force–velocity abilities, which are more related to anaerobic capacity and neuromuscular factors [7,20–22]. Furthermore, this bio-energetic shift necessitates ulterior technical adjustments such as maintaining stroke efficiency (i.e., sustaining parameters of stroke length, stroke frequency, and stroke index) for a longer time compared to 50 m and 100 m swimming races [34,35]. The generally lower force–velocity values in middle-distance swimmers may be also favored by the greater configuration of technical parameters from a tactical–strategical perspective, which is less present in sprinting competitions [34].

Still, there are several reasons to advocate for a leveling up of maximum force–velocity levels even in middle-distance swimmers competing at the regional level. For instance, in the present study, we found a considerable correlation between higher productions of force–velocity and superior swimming performances, including the 200 m performers. Moreover, this is in line with several scholars who observed that underdeveloped levels of force–velocity can result in an early deterioration of technical skills due to the accumulation of neuromuscular fatigue [7]. These aspects would also definitely benefit the in-water performance of middle-distance swimmers. Therefore, we strongly encourage trainers to fill the apparent gap in bio-motor skills between sprinters and middle-distance swimmers, providing the latter with more focus and training time to upgrade their force–velocity capacities.

In addition, possible alterations in maximum force–velocity production due to specific training periods should be considered [5]. For instance, in this study, we collected force–velocity data at the end of the swimmers' preparatory cycle of training (i.e., after the first 8 weeks of training). However, considering both the differences in training as well as the significant gap in bio-motor abilities that we found between sprinters and 200 m performers, it may be that the force–velocity capacities of middle-distance swimmers are greatly susceptible to the variations in training intensity and volume occurring over the season (e.g., from the preparatory cycle of training to the competitive cycle of training). For these reasons, we recommend future studies to carefully analyze hypothetical fluctuations in swimmers' force–velocity levels over a competitive season and how these fluctuations may affect swimming performances, especially for middle-distance swimmers.

Notably, we found a significant gap in maximum force–velocity production in breaststroke sprinters compared to the other 50 m and 100 m strokes (Figure 2). Indeed, we should account for some technical and biomechanical restraints regarding stroke velocity and general efficiency in breaststrokes, especially compared to the butterfly, backstroke, and front-crawl styles of swimming. Here, the basic assumption is that in order to reach, maintain, and increase in-water velocity, swimmers must continuously generate muscular propulsive forces to "fight" and exceed the drag forces of water. However, it is worth mentioning that breaststroke swimming produces the largest intracycle velocity variability among the four strokes [34]. This is due to the added drag of recovering both arms under the water and in drawing the knees up to prepare for the next propulsive phase of the stroke cycle. In fact, breaststroke is the sole stroke that does not contemplate the arm-pushing phase. Instead, it is the lower body that is responsible for the active propulsive phase during the stroke. Moreover, it has been shown that the energy expenditure during butterfly and breaststroke swimming is approximately twofold greater than in backstroke or front-crawl swimming [34]. Again, this was due to the increase in form drag dictated by the mechanics of these strokes. However, despite both butterfly and breaststroke sharing a symmetrical movement pattern, the breaststroke was shown to be the least efficient stroke in terms of energy expenditure and general in-water velocity. In fact, Pyne et al. [34] observed that the front crawl presented the lowest energy cost (1.23 kJ/m^{-1}), followed by backstroke (1.47 kJ/m^{-1}), butterfly (1.55 kJ/m^{-1}), and breaststroke (1.87 kJ/m^{-1}). Moreover, the swimming energy cost increased exponentially with an increase in swim velocity during freestyle, backstroke, and butterfly, but this change was linear in breaststroke [34].

In this regard, our findings transpose the in-water biomechanical disadvantages of breaststroke specialists into dry-land bio-motor disparities. The apparent bio-motor limitations on dry-land, the higher complexity of neuromuscular coordination between upper and lower limbs, as well as the inferior mechanical efficiency, put breaststroke in a unique as well as critical position regarding specific performance evaluation and improvement. All of the evidence considered, it may be that breaststroke performers depend more on maximizing their technical ability instead of their force–velocity production in a vertical pulling motion. In addition, given the major involvement of the lower body in generating propulsive forces during breaststroke, it is possible that the different contributions of the legs would be reflected in different force–velocity exertions between breaststroke and the other three strokes. In particular, we would suggest testing this hypothesis using either the back squat or bodyweight vertical jumps (e.g., countermovement jump), which are the most used and effective motions for indirectly improving the "lower-body-focused" elements of swimming races (i.e., diving and turning) [11,18,21].

This study is not exempt from limitations. Namely, the subjects enrolled had very specific characteristics regarding their competitive level (regional), training experience (6.5 ± 1.1 years of experience), and gender (male). On the one hand, the sample homogeneity allowed us to thoroughly analyze and compare several aspects of maximum force–velocity exertion and swimming performances. On the other hand, we cannot state if the findings from the present study would be confirmed either in athletes competing at the national/international level, holding more years of experience, or considering a population of female swimmers. Moreover, we only recruited swimmers specialized in a single stroke; however, swimmers can often compete over multiple specialties or medleys. What would the force–velocity capacities of this multi-specialized athlete be like? Perhaps, the higher grade of cross-training among strokes could bring some sort of technical/bio-motor transfer, which trainers should purposely take advantage of in order to improve specific aspects of a single stroke. However, this speculation needs to be verified with apt experimental designs investigating possible changes in swimmers' maximum force–velocity exertion due to multifaceted training–competitive approaches.

5. Conclusions

Measuring maximum force–velocity exertion with the single pull-up test in regional-level swimmers may be a plain, scalable, lab-independent, cost-effective, and time-efficient experimental approach, apparently capable of discerning different levels of neuromuscular abilities based on stroke and distance specialization. However, it is debatable whether the results provided in this study are indeed a manifestation of different degrees of force–velocity capacities among distinctive categories of specialized swimmers, especially between sprinters and middle-distance swimmers, and between breaststroke and the other strokes. Therefore, we encourage continued investigation into this topic, to inform the process of developing evidence-based recommendations for scholars and trainers interested in enhancing swimming performance.

Finally, other sports could benefit from the evaluation of maximum force–velocity exertion for performance prediction and differentiation, especially closed-skill ones (as in swimming). This is because these kinds of sporting activities present almost no peer interactions and few environmental elements capable of affecting athletic performance, thus conceding sheer bio-motor abilities with considerable clout on the competitive outcome. However, it is also worth considering that swimming possesses many environmental elements that can affect performance and that differentiate it from other sports that are practiced on land in contrast to water. For these reasons, while the framework we proposed in this article could be incorporated within other sporting environments, it should also be rearranged for the specific sporting activity, with the ultimate goal of assessing and optimizing athletic performance for competitive endeavors.

Author Contributions: Conceptualization, V.S., J.M.G.R., A.A.-O. and A.L.-H.; methodology, V.S., A.A.-O. and J.M.G.R.; software, V.S. and J.S.d.C.; validation, D.M. and J.M.G.R.; formal analysis, J.S.d.C.; investigation, V.S.; A.A.-O. and A.L.-H.; resources, J.M.G.R.; data curation, V.S. and J.S.d.C. writing—original draft preparation, V.S.; writing—review and editing, J.M.G.R.; visualization, V.S., A.A.-O. and D.M.; supervision, J.M.G.R.; project administration, J.M.G.R. All authors have read and agreed to the published version of the manuscript.

Funding: This research received no external funding.

Institutional Review Board Statement: The study was approved by the local Ethics Committee of the university (FGM02102019) and was conducted in accordance with the Declaration of Helsinki.

Informed Consent Statement: Informed consent was obtained from all of the subjects involved in the study.

Data Availability Statement: Data are available on request.

Conflicts of Interest: The authors declare no conflict of interest.

References

1. Wang, C.-H.; Chang, C.-C.; Liang, Y.-M.; Shih, C.-M.; Chiu, W.-S.; Tseng, P.; Hung, D.L.; Tzeng, O.J.L.; Muggleton, N.G.; Juan, C.-H. Open vs. closed skill sports and the modulation of inhibitory control. *PLoS ONE* **2013**, *8*, e55773. [CrossRef] [PubMed]
2. Sadowski, J.; Mastalerz, A.; Gromisz, W. Transfer of dry-land resistance training modalities to swimming performance. *J. Hum. Kinet.* **2020**, *74*, 195–203. [CrossRef]
3. Amaro, N.M.; Marinho, D.A.; Marques, M.C.; Batalha, N.P.; Morouço, P.G. Effects of Dry-Land Strength and Conditioning Programs in Age Group Swimmers. *J. Strength Cond. Res.* **2017**, *31*, 2447–2454. [CrossRef] [PubMed]
4. Crowley, E.; Harrison, A.J.; Lyons, M. The impact of resistance training on swimming performance: A systematic review. *Sport. Med.* **2017**, *47*, 2285–2307. [CrossRef] [PubMed]
5. Mujika, I.; Halson, S.; Burke, L.M.; Balagué, G.; Farrow, D. An integrated, multifactorial approach to periodization for optimal performance in individual and team sports. *Int. J. Sport. Physiol. Perform.* **2018**, *13*, 538–561. [CrossRef]
6. Muniz-Pardos, B.; Gomez-Bruton, A.; Matute-Llorente, A.; Gonzalez-Aguero, A.; Gomez-Cabello, A.; Gonzalo-Skok, O.; Casajus, J.A.; Vicente-Rodriguez, G. Swim-specific resistance training: A systematic review. *J. Strength Cond. Res.* **2019**, *33*, 2875–2881. [CrossRef]
7. Bompa, T.O.; Buzzichelli, C.A. *Periodization: Theory and Methodology of Training*, 6th ed.; Human Kinetics: Champaign, IL, USA, 2019.
8. Komi, P.V. Strength and power in sport. *Med. Sci. Sport. Exerc.* **1994**, *26*, 1422. [CrossRef]
9. Lum, D.; Barbosa, T.M. Effects of strength training on olympic time-based sport performance: A systematic review and meta-analysis of randomized controlled trials. *Int. J. Sport. Physiol. Perform.* **2019**, *14*, 1318–1330. [CrossRef]
10. Berryman, N.; Mujika, I.; Bosquet, L. Concurrent training for sports performance: The 2 sides of the medal. *Int. J. Sport. Physiol. Perform.* **2019**, *14*, 279–285. [CrossRef]
11. Crowley, E.; Harrison, A.J.; Lyons, M. Dry-land resistance training practices of elite swimming strength and conditioning coaches. *J. Strength Cond. Res.* **2018**, *32*, 2592–2600. [CrossRef]
12. Duchateau, J.; Stragier, S.; Baudry, S.; Carpentier, A. Strength training: In search of optimal strategies to maximize neuromuscular performance. *Exerc. Sport Sci. Rev.* **2021**, *49*, 2–14. [CrossRef] [PubMed]
13. Suchomel, T.J.; Nimphius, S.; Bellon, C.R.; Stone, M.H. The importance of muscular strength: Training considerations. *Sport. Med.* **2018**, *48*, 765–785. [CrossRef] [PubMed]
14. Ikuta, Y.; Matsuda, Y.; Yamada, Y.; Kida, N.; Oda, S. Relationship between changes of swimming velocity, stroke rate, stroke length and muscle activities in front crawl swimming. *Jpn. J. Phys. Fit. Sport. Med.* **2010**, *59*, 427–438. [CrossRef]
15. Costa, M.J.; Balasekaran, G.; Vilas-Boas, J.P.; Barbosa, T.M. Physiological adaptations to training in competitive swimming: A systematic review. *J. Hum. Kinet.* **2015**, *49*, 179–194. [CrossRef]
16. Gonjo, T.; Njos, N.; Eriksrud, O.; Olstad, B.H. The relationship between selected load-velocity profile parameters and 50 m front crawl swimming performance. *Front. Physiol.* **2021**, *12*. [CrossRef]
17. Pleša, J.; Kozinc, Ž.; Šarabon, N. A brief review of selected biomechanical variables for sport performance monitoring and training optimization. *Appl. Mech.* **2022**, *3*, 144–159. [CrossRef]
18. Born, D.-P.; Stöggl, T.; Petrov, A.; Burkhardt, D.; Lüthy, F.; Romann, M. Analysis of freestyle swimming sprint start performance after maximal strength or vertical jump training in competitive female and male junior swimmers. *J. Strength Cond. Res.* **2020**, *34*, 323–331. [CrossRef]
19. Lopes, T.J.; Neiva, H.P.; Gonçalves, C.A.; Nunes, C.; Marinho, D.A. The effects of dry-land strength training on competitive sprinter swimmers. *J. Exerc. Sci. Fit.* **2021**, *19*, 32–39. [CrossRef]
20. Carvalho, D.; Soares, S.; Zacca, R.; Marinho, D.; Silva, A.; Pyne, D.; Vilas-Boas, J.; Fernandes, R. In-water and on-land swimmers' symmetry and force production. *Int. J. Environ. Res. Public Health* **2019**, *16*, 5018. [CrossRef]

21. Pérez-Olea, J.I.; Valenzuela, P.L.; Aponte, C.; Izquierdo, M. Relationship between dryland strength and swimming performance: Pull-up mechanics as a predictor of swimming speed. *J. Strength Cond. Res.* **2018**, *32*, 1637–1642. [CrossRef]
22. Hermosilla, F.; Sanders, R.; González-Mohíno, F.; Yustres, I.; González-Rave, J.M. Effects of dry-land training programs on swimming turn performance: A systematic review. *Int. J. Environ. Res. Public Health* **2021**, *18*, 9340. [CrossRef] [PubMed]
23. Morais, J.E.; Silva, A.J.; Garrido, N.D.; Marinho, D.A.; Barbosa, T.M. The transfer of strength and power into the stroke biomechanics of young swimmers over a 34-week period. *Eur. J. Sport Sci.* **2018**, *18*, 787–795. [CrossRef]
24. Garrido, N.; Marinho, D.A.; Reis, V.M.; van den Tillaar, R.; Costa, A.M.; Silva, A.J.; Marques, M.C. Does combined dry land strength and aerobic training inhibit performance of young competitive swimmers? *J. Sport. Sci. Med.* **2010**, *9*, 300–310.
25. Hermosilla, F.; Yustres, I.; Psycharakis, S.; Santos del Cerro, J.; González-Mohíno, F.; González-Rave, J.M. Which variables may affect underwater glide performance after a swimming start? *Eur. J. Sport Sci.* **2022**, *22*, 1141–1148. [CrossRef]
26. Dundar, A.; Kocahan, S.; Arslan, C. Effects of different loading exercises on apelin levels and physical and hematologic parameters of swimmers. *Horm. Mol. Biol. Clin. Investig.* **2019**, *38*. [CrossRef] [PubMed]
27. Martens, J.; Figueiredo, P.; Daly, D. Electromyography in the four competitive swimming strokes: A systematic review. *J. Electromyogr. Kinesiol.* **2015**, *25*, 273–291. [CrossRef]
28. Gonjo, T.; Narita, K.; McCabe, C.; Fernandes, R.J.; Vilas-Boas, J.P.; Takagi, H.; Sanders, R. Front crawl is more efficient and has smaller active drag than backstroke swimming: Kinematic and kinetic comparison between the two techniques at the same swimming speeds. *Front. Bioeng. Biotechnol.* **2020**, *8*, 570657. [CrossRef]
29. Aspenes, S.T.; Karlsen, T. Exercise-training intervention studies in competitive swimming. *Sport. Med.* **2012**, *42*, 527–543. [CrossRef]
30. Kavanaugh, A. The role of progressive overload in sports conditioning. *Strength Cond. J.* **2008**, *6*, 15–17.
31. González-Ravé, J.M.; Pyne, D.B.; Castillo, J.; González-Mohíno, F.; Stone, M. Training periodization for a world-class 400 meters individual medley swimmer. *Biol. Sport* **2022**, *39*, 883–888. [CrossRef]
32. González Ravé, J.M. The traditional periodization in individual sports: Providing effective responses to both new and old problems. *Arch. Med. Deporte* **2021**, *38*, 76–77. [CrossRef]
33. Kraemer, W.J.; Fleck, S.J. *Optimizing Strength Training: Designing Nonlinear Periodization Workouts*; Human Kinetics: Champaign, IL, USA, 2007.
34. Pyne, D.B.; Sharp, R.L. Physical and energy requirements of competitive swimming events. *Int. J. Sport Nutr. Exerc. Metab.* **2014**, *24*, 351–359. [CrossRef] [PubMed]
35. Barbosa, T.; Fernandes, R.; Keskinen, K.; Colaço, P.; Cardoso, C.; Silva, J.; Vilas-Boas, J. Evaluation of the energy expenditure in competitive swimming strokes. *Int. J. Sport. Med.* **2006**, *27*, 894–899. [CrossRef] [PubMed]

Journal of
Functional Morphology and Kinesiology

MDPI

Article

Pivot Step Jump: A New Test for Evaluating Jumping Ability in Young Basketball Players

Apostolos S. Theodorou [1,*], Hariklia-Parthenia Rizou [1], Emmanouil Zacharakis [1], Ioannis Ktistakis [1], Evangelos Bekris [1], Vassilios Panoutsakopoulos [2], Panagiotis Strouzas [1], Dimitrios I. Bourdas [1] and Nikolaos Kostopoulos [1]

[1] School of Physical Education & Sport Sciences (TEFAA), National and Kapodistrian University of Athens, Dafni, 17237 Athens, Greece
[2] Biomechanics Laboratory, School of Physical Education and Sports Sciences at Thessaloniki, Aristotle University of Thessaloniki, 54124 Thessaloniki, Greece
* Correspondence: aptheod@phed.uoa.gr; Tel.: +30-210-727-6175

Abstract: Jumping ability in basketball is usually assessed using standardized vertical jump tests. However, they lack specificity and do not consider the player's basketball skills. Several studies have suggested performing specific jump tests, which are tailored to the movement patterns and requirements of a basketball game. The pivot step jump test (PSJT) is a novel test designed to evaluate the specific jumping abilities of basketball players by combining a pivot step on one leg with a maximum bilateral vertical jump. This study had two aims: to determine the reliability and validity of the PSJT using typical jump tests as the criterion measure and to demonstrate the PSJT as a practical test to evaluate specific jumping ability in young male and female basketball players. Twenty female (EGA; 14.0 ± 0.7 years, 59.3 ± 7.9 kg, 162.1 ± 5.5 cm) and fifteen male (EGB; 14.0 ± 0.7 years, 58.1 ± 7.7 kg, 170.3 ± 6.4 cm) basketball players participated in the study. The test–retest reliability of the PSJT within sessions (intrasession reliability) and across sessions (intersession reliability) was assessed within EGA. For the evaluation of validity, EGB performed the PSJT and a series of criterion jumping tests. For EGA, no changes ($p > 0.05$) were found in PSJT performance between test sessions and excellent intra- and intersession reliability was observed (ICCs > 0.75). Correlation coefficients indicated high factorial validity between the jumping tests and PSJT ($r = 0.71$–0.91, $p < 0.001$). The PSJT appears to offer a valid assessment of jumping ability in basketball and is a practical test for assessing sport-specific jumping skills in young basketball players.

Keywords: vertical jump; interlimb asymmetry; performance; power; sport-specific skill; developmental age; motor skill assessment

Citation: Theodorou, A.S.; Rizou, H.-P.; Zacharakis, E.; Ktistakis, I.; Bekris, E.; Panoutsakopoulos, V.; Strouzas, P.; Bourdas, D.I.; Kostopoulos, N. Pivot Step Jump: A New Test for Evaluating Jumping Ability in Young Basketball Players. *J. Funct. Morphol. Kinesiol.* **2022**, *7*, 116. https://doi.org/10.3390/jfmk7040116

Academic Editor: Diego Minciacchi

Received: 14 November 2022
Accepted: 8 December 2022
Published: 15 December 2022

Publisher's Note: MDPI stays neutral with regard to jurisdictional claims in published maps and institutional affiliations.

1. Introduction

Basketball is a physically demanding sport where success dependents on a variety of fundamental physical skills such as acceleration, quickness, strength, and power [1,2]. During a basketball game, bilateral and unilateral jumps are performed at the frontal and sagittal plane of motion. However, most training programs emphasize drills, which are performed at the sagittal plane and rarely examine the effects of training at the other planes of motion. The specialization of training implies that fitness assessment should include actions that are kinematically similar to the movements of a given sport. Jumping is one of the basic actions performed during a basketball game, as the basket is at a height of 3.05 m [3]. Jumping ability is a manifestation of power, which is a sport-specific feature of basketball [4]. All actions involving jumps are affected by a range of different factors pertinent to the game of basketball [2]. A large variety of the offensive and defensive skills used in the sport, such as shots, lay ups, rebounds, etc., involve jumping. Jumping ability, therefore, determines the performance and level of basketball players [2] and should be adapted to the specific requirements the game.

Jumping ability in basketball is widely assessed with a wide range of standardized jumping tests [5], i.e., the squat jump, countermovement jumps with or without arm swing, the drop jump, and the standing long jump [3,6,7]. Performance in these tests assesses the players' ability to utilize the stretch shortening cycle and/or arm swing to increase jump height [8]. In the case of the countermovement jump, its execution with (CMJA) or without arm swing (CMJ) is suggested to be reliable and valid in measuring peak power of lower limbs and jump height [9,10]. Furthermore, the difference between CMJ and CMJA provides an indication of neuromuscular ability to coordinate intra-segmental energy flow to achieve higher power and jump height [11–13]. However, basketball players differ from other athletes in terms of both the magnitude of power output and the time instance of its peak, as they execute the vertical jumps in a sport-specific force- and time-dependent pattern [14–20]. It can be argued that typical vertical jump tests are generic regarding the performance evaluation of basketball players since they lack specificity and do not consider the player's basketball skills. For example, in basketball, players perform a variety of different types of jumps, i.e., jumps from a still position and jumps following a running action [21]. These jumps can be influenced by different factors related to the game, such as the path of the ball, physical contact with opponents and the phase of the game, offense, or defense [22]. Several studies have questioned the validity of these tests in assessing the functional abilities of basketball players and have suggested that specific jump tests tailored to movement patterns and the demands of a basketball game be performed [22,23]. In addition, it has been suggested that performing vertical jump tests on the court is more appropriate and appealing for basketball players when evaluating their jumping ability [10]. In addition, basketball-specific physical field tests are important for monitoring training effectiveness and fitness status [24]. In the case of jumping, the specific jump tests used in basketball are the "three-steps approach with two leg take-off vertical jump" test, the "two-steps approach with one leg take-off vertical jump" test [2], and the "one-step jump" test [25]. The concept underlying the aforementioned jump tests is to assess specific aspects of basketball players' strength and conditioning abilities by combining jumping with game-specific skills that involve a jumping action.

To be effective in the game and utilize his jumping ability, a player must adapt this ability to the context and specific requirements of the sport. Physical qualities and tactics are currently considered as two inseparable representations of a player's actions, and therefore it is crucial to take a more ecological approach when training and evaluating athletes [26]. For basketball players to be successful, they must be able to carry out multiple power-based actions before jumping such as cutting and dribbling simultaneously. Therefore, measuring vertically oriented power-related attributes requires a targeted approach that the current power-related tests perform with limited ecological validity. The present study proposes a novel jump test for a functional assessment of basketball jumping abilities. Pivoting is when a player stands still and steps with one foot. The foot that stays on the ground is called the pivot foot. A player that has the ball and is standing still may step with one foot in order to change direction and pass the ball or to avoid opponents and take a shot. Pivots and jumps are combined in all these actions. The pivot step jump is a step on one leg (while the other remains in contact with the ground) for changing direction, followed by a maximum vertical jump on both feet. Basketball players usually perform this movement after rebounds and to find a better position for a shot when blocked by the opponents [3]. The jump following a pivot has never been studied and evaluated as a test. Considering that talent identification and long-term development require the inclusion of tests of technical ability and tactical behavior [27], the purpose of this study was to investigate the validity and reliability of a new basketball jump test, that involves a vertical jump following a pivot action. The test was referred to as the pivot step jump test (PSJT). We hypothesized that the PSJT would have (a) high intrasession, intersession and interrater reliability and (b) a strong relation with the vertical jump tests most used to evaluate power in basketball players.

2. Materials and Methods

2.1. Experimental Approach to the Problem

The present study was designed to determine the concurrent reliability and validity of a new jump test by examining correlations with established and previously validated jump tests. Concurrent validity is a type of criterion-related validity in which correlation coefficients are calculated between a true criterion and an alternative measure. Previous research has proposed using this method to assess the validity of upper-body power tests [28] and lower-body power [29]. To test the reliability and validity of the proposed jump test, each participant performed the test in four instances. To avoid a possible training effect due to the athletes' prolonged and repeated participation in tests with maximal effort, two groups of participants were tested, with each group assigned to test each hypothesis.

2.2. Participants

Fifteen male (EGA; age: 14.0 ± 0.65 years) and twenty female basketball players (EGB, aged 14.0 ± 0.65 years) participated at the study. The anthropometric characteristics of the participants are shown in Table 1. All athletes had at least 5 years of experience in basketball and competed in the first division of their age group. All experimental procedures were approved by the Institutional Research Ethics and Bioethics Committee (1063/13 June 2018). All participants and their guardians were informed about the benefits and risks of the study. Signed parental consent was obtained for the participation of minor athletes.

Table 1. Anthropometric characteristics of the participants (mean ± SD).

Group	Body Height (cm)	Body Mass (kg)	Body Mass Index (kg/m^2)
EGA (males; n = 15)	170.26 ± 6.43	58.13 ± 7.69	20.05 ± 2.89
EGB (females; n = 20)	162.07 ± 5.48	59.29 ± 7.87	22.57 ± 2.63

EGA: Experimental group A; EGB: Experimental group B.

2.3. Experimental Procedure

All testing was conducted in the off-season to avoid the effects of team training. EGA group was used to determine test–retest reliability of the new jump test within sessions (intrasession reliability), within investigators (interrater reliability) and across sessions (intersession reliability). EGB group was used to determine the validity of the new test with four traditional jump tests. Participants at the EGA group reported to the laboratory on four separate occasions (Figure 1).

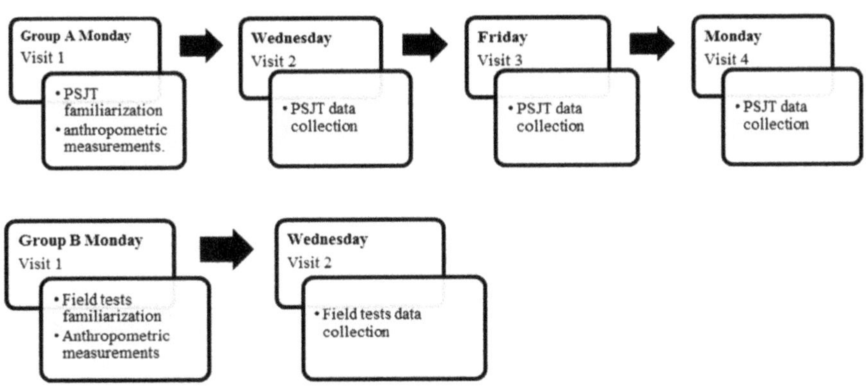

Figure 1. Graphical representation of the design of the study for group A (EGA) and group B (EGB).

On the first visit, participants familiarized themselves with the PSJT and took anthropometric measurements (mass, standing height, seated height, leg length, and shin length). Measurements were performed according to Carter and Heath [30], using the Seca 220 telescopic measuring rod, the Seca Alpha 770 scale and Seca 201 measurement tape (Seca GmbH & Co., Hamburg, Germany). The anthropometric data were used to estimate the maturity level of the participants by using the simplified regression equations for maturity adjustment proposed by Moore et al. [31].

Prior to testing, all participants completed an 8-min warm-up consisting of submaximal plyometric and jump drills. During the familiarization period, detailed instructions were provided on how to execute the PSJT and 4 trials were performed with emphasis on proper execution technique.

The PSJT was administered using the Optogait system (Microgate, Bolzano, Italy). Participants commenced from a stationary semi-squat position (knee angle of 90°), with one leg (right, $PSJT_{RLEG}$ or left, $PSJT_{LLEG}$) inside the Optogait's measuring rods. Then, they performed a forward 90° pivot step followed by a rapid vertical jump, using an arm swing aiming to reach maximum height. During the PSJT, participants were allowed to perform a countermovement of the legs and arms (Figure 2).

Figure 2. Graphical representation of the design of the study.

The height of the jump was recorded as the result of the test. Three attempts were allowed for each leg. The mean of the two best attempts was used for analysis. The same procedure was then repeated for the other leg. Following a two-day rest period (Visit 2), participants performed the PSJT twice (tests 1 and 2), separated by a 30 min interval, under the supervision of investigator "A". A third PSJT was performed two days later (Visit 3), under the supervision of investigator "B", and a fourth PSJT was performed seven days later (Visit 4), again under the supervision of investigator "B".

EGB participants reported to the laboratory on two separate occasions to familiarize themselves with the testing procedure and data collection. On the first visit, participants completed a familiarization with the PSJT and the four jump tests and took anthropometric measurements. After two days (Visit 2), participants repeated the five jump tests. The countermovement jump (CMJ) commenced from a stationary upright standing position (hands akimbo) followed by a preliminary downward movement by flexing the knees and hips and subsequently vigorously extending them to perform a vertical jump [32]. The countermovement jump with arm swing (CMJA) was conducted following the same procedure as the CMJ, but participants were allowed to perform the jump with an arm swing. The effect of arm swing on countermovement was determined using the arm swing augmentation index (AS_{INDEX}), which was calculated as the percentage ratio of the difference in jump height between the CMJA and CMJ divided by the jump height in CMJ. The countermovement jumps with the right or left leg (CMJ_{RLEG} and CMJ_{LLEG}) were performed similarly to the CMJA, but with one leg. Performance in all jump tests was evaluated by jump height, estimated from the time of flight measured by the Optogait system [22,33]. To investigate the presence of possible asymmetry between CMJ_{RLEG} and CMJ_{LLEG}, and between $PSJT_{RLEG}$ and $PSJT_{LLEG}$, the respective asymmetry values were quantified based on the symmetry angle (θ_{SYM}) [34].

An additional jump test to assess lower extremities explosive strength, was the standing long jump (SLJ). The SLJ commenced with the toes behind a take-off line. By bending the knees and swinging the arms freely, the participant performed a horizontal jump to cover the greatest horizontal distance possible. The distance was measured from the take-off line to the rearmost heel [35]. The jump tests were performed in a random order. For each test, three attempts were allowed with a 1-min rest period between attempts and 8 min between tests. The mean of the two best attempts was used for analysis.

2.4. Statistical Analysis

Statistical analysis for this study was performed using IBM SPSS Statistics v.27.0.1.0 (IBM Corp., Armonk, NY, USA) software. The level of significance was set at $a = 0.05$. After assessing the normality of the data (each test session for each leg) with the Kolmogorov–Smirnov test, the means and standard deviations for all variables were calculated. The p values obtained by the Kolmogorov–Smirnov analyses were all above 0.05, indicating that the data were normally distributed. The intrasession (test 1 vs. test 2), interrater (test 2 vs. test 3), and intersession (test 3 vs. test 4) reliability of the PSJT measures was quantitatively assessed with two-way random, single measure intraclass correlation coefficient (ICC) and their respective 95% CI. An ICC with values >0.75, ≥0.40 and ≤0.75, and <0.40 indicated "excellent", "fair to good", and "poor" reliability, respectively [36]. In addition, Bland–Altman plots were used to determine the extent of agreement between test–retest values. The difference in the paired intrasession, interrater and intersession measures was plotted against their respective means. The evaluation criterion was that 95% of the data points should lie within the mean ± 2 SDs of the differences for the intra and intersession measurements, which corresponds to the 95% CI. Absolute reliability was calculated using the standard error of measurement (SEM) [37], which was then expressed as a percentage of the mean value with the coefficient of variation (CoV). A CoV value less than 10% was set as a criterion for an acceptable reliability [22]. The minimal difference (MD) in absolute terms (cm) and as a percentage of the mean value (MD%) was also determined [37]. The factorial validity of the jump tests and the relationships between the measured variables were determined using Pearson's correlation coefficients.

3. Results

The seated height and the lengths of the body segments measured are shown in Table 2. The maturity offset was found to be positive in both groups.

Table 2. Mean ± SD of the anthropometric measurements (n = 15).

Group	Seated Height (cm)	Leg Length (cm)	Shin Length (cm)	Maturity Offset (yrs)
EGA (males; n = 15)	87.57 ± 3.47	110.14 ± 4.21	45.46 ± 2.82	0.9 ± 0.4
EGB (females; n = 20)	84.95 ± 3.51	102.42 ± 4.05	41.67 ± 1.65	2.7 ± 0.3

EGA: experimental group A; EGB: experimental group B.

3.1. PSJT Reliability

Descriptive statistics of the testing sessions for EGA are provided in Table 3. The calculated ICC and 95% CI values for the PSJT performances across the three reliability analyses are presented in Table 4. The ICCs were all above 0.75, indicating an excellent intra and intersession reliability.

Table 3. Mean ± SD of the PSJT measures obtained during all testing sessions for EGA (n = 15).

Test	$PSJT_{LLEG}$ (cm)	$PSJT_{RLEG}$ (cm)
Test 1	37.31 ± 6.09	36.97 ± 6.24
Test 2	37.40 ± 5.71	37.02 ± 5.82
Test 3	37.27 ± 5.90	37.12 ± 6.32
Test 4	37.07 ± 5.82	36.91 ± 6.05

$PSJT_{LLEG}$: pivot step jump test on the left leg; $PSJT_{RLEG}$: pivot step jump test on the right leg.

Table 4. Interclass correlation coefficients (ICC), 95% confidence interval (95% CI), standard error of measurement (SEM), coefficient of variation (CoV), and minimal difference (MD) values of the intrasession, interrater, and intersession reliability analysis of the PSJT measures for EGA (n = 15).

Reliability Analysis	Intra-Session		Inter-Rater		Inter-Session	
	$PSJT_{LLEG}$	$PSJT_{RLEG}$	$PSJT_{LLEG}$	$PSJT_{RLEG}$	$PSJT_{LLEG}$	$PSJT_{RLEG}$
ICC	0.992 *	0.981 *	0.98 5 *	0.978 *	0.991 *	0.983 *
95% CI	0.975–0.997	0.946–0.994	0.956–0.995	0.970–0.993	0.975–0.987	0.950–0.994
SEM (cm)	0.694	0.810	0.694	0.876	0.544	0.786
CoV (%)	1.384	2.190	1.859	2.364	1.463	2.124
MD (cm)	2.541	3.981	3.41	4.307	2.673	3.863
MD (%)	6.803	10.759	9.135	11.618	7.191	10.438

*: $p < 0.01$; $PSJTL_{LEG}$: pivot step jump test on the left leg; $PSJT_{RLEG}$: pivot step jump test on the right leg.

Bland–Altman plots and the regression analyses results for the PSJT performances for the intra- and inter-session comparisons are shown in Figures 3–5 and Table 5. For all comparisons, data points were within the mean ± 2 SD.

Table 5. Interclass correlation coefficients (ICC), 95% confidence interval (95% CI), standard error of measurement (SEM), coefficient of variation (CoV), and minimal difference (MD) values of the intrasession, interrater and intersession reliability analysis of the PSJT measures for EGA (n = 15).

Comparisons	F	p	R^2 Adjusted
Intrasession Test1 vs. Test2—$PSJT_{LLEG}$	(1,13) = 926.05	<0.001	0.986
Intrasession Test1 vs. Test2—$PSJT_{RLEG}$	(1,13) = 360.06	<0.001	0.963
Interrater Test2 vs. Test3—$PSJT_{LLEG}$	(1,13) = 406.73	<0.001	0.967
Interrater Test2 vs. Test3—$PSJT_{RLEG}$	(1,13) = 319.17	<0.001	0.958
Intersession Test3 vs. Test4—$PSJT_{LLEG}$	(1,13) = 731.99	<0.001	0.981
Intersession Test3 vs. Test4—$PSJT_{RLEG}$	(1,13) = 370.75	<0.001	0.964

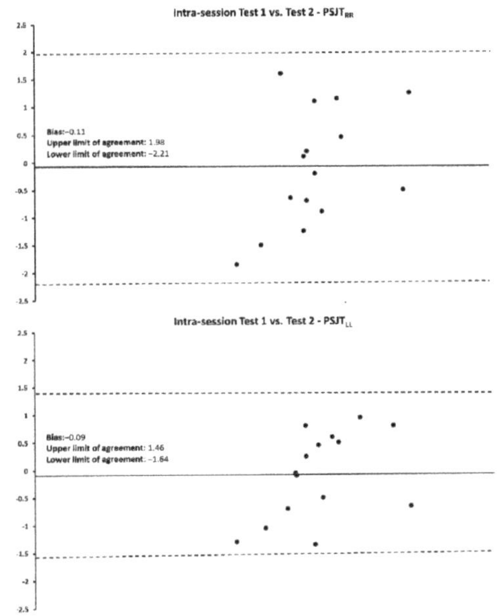

Figure 3. Bland–Altman plot for the intrasession reliability between PSJT$_{RLEG}$ and PSJT$_{LLEG}$ for EGA. The difference between the two measurements per subject is plotted against the mean of the two measurements. The dotted lines represent the upper and lower limits of agreement for 95% CI.

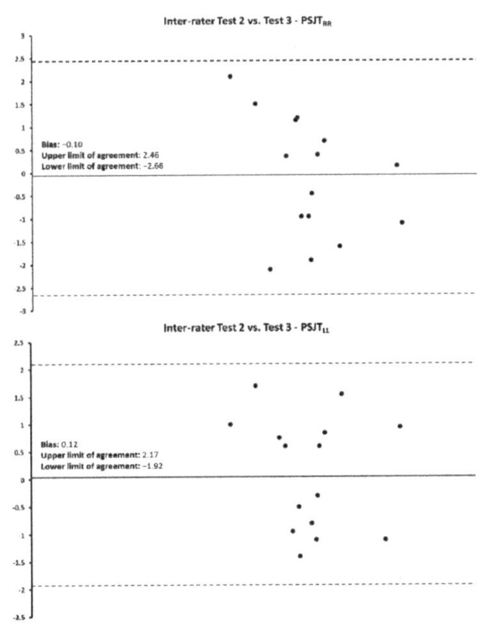

Figure 4. Bland–Altman plot for the interrater reliability between PSJT$_{RLEG}$ and PSJT$_{LLEG}$ for EGA. The difference between the two measurements per subject is plotted against the mean of the two measurements. Dotted lines represent the upper and lower limits of agreement for 95% CI.

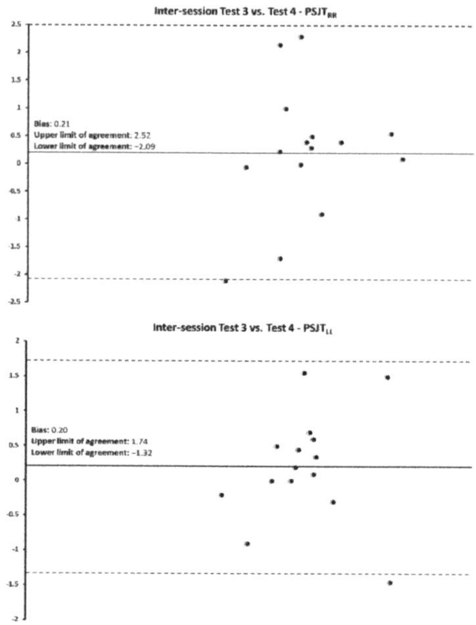

Figure 5. Bland–Altman plot for the intersession reliability between $PSJT_{RLEG}$ and $PSJT_{LLEG}$ for EGA. The difference between the two measurements per subject is plotted against the mean of the two measurements. The dotted lines represent the upper and lower limits of agreement for 95% CI.

3.2. PSJT Validity

Table 6 shows the results of all jumping tests examined. No significant interlimb asymmetry was observed in both the CMJ and the PSJT.

Table 6. Descriptive statistics for all tests and indexes examined in the present study for EGB (females, $n = 20$).

Test	Mean ± SD	95% CI
CMJ (cm)	23.04 ± 4.08	21.13–24.95
CMJA (cm)	26.64 ± 4.61	24.48–28.80
AS_{INDEX} (%)	15.91 ± 7.24	12.51–19.29
CMJ_{RLEG} (cm)	11.48 ± 2.45	10.33–12.62
CMJ_{LLEG} (cm)	11.00 ± 2.39	9.87–12.12
$\theta_{SYM\text{-}CMJ}$ (deg)	−1.46 ± 4.53	−3.57–0.66
$PSJT_{RLEG}$ (cm)	25.46 ± 4.42	23.39–27.53
$PSJT_{LLEG}$ (cm)	25.22 ± 4.65	23.04–27.40
$\theta_{SYM\text{-}PSJT}$ (deg)	−0.35 ± 2.40	−1.47–0.77
SLJ (cm)	155.50 ± 25.47	143.57–167.42

CMJ: bilateral countermovement jump—arms akimbo; CMJA: bilateral countermovement jump with an arm swing; AS_{INDEX}: arm swing augmentation index; CMJ_{RLEG}: unilateral countermovement jump performed with the right leg; CMJ_{LLEG}: unilateral countermovement jump performed with the left leg; $\theta_{SYM\text{-}CMJ}$: symmetry angle for the unilateral countermovement jumps; $PSJT_{RLEG}$: pivot step jump performed with the right leg; $PSJT_{LLEG}$: pivot step jump performed with the left leg; $\theta_{SYM\text{-}PSJT}$: symmetry angle for the pivot step jump tests; SLJ: standing long jump.

All correlation coefficients indicated significant correlations among the jump tests ($p < 0.05$). When univariate associations between variables were examined, correlations between jump tests and PSJT performances ranged from 0.71 to 0.91, indicating that the jump tests assessed and the PSJT have high factorial validity (Table 7).

Table 7. Intercorrelation matrix of the PSJT and the other jump tests for EGB (females, $n = 20$).

Test	CMJ r (p)	CMJA r (p)	CMJ_{RLEG} r (p)	CMJ_{LLEG} r (p)	SLJ r (p)	PSJ_{RLEG} r (p)	PSJ_{LLEG} r (p)
CMJ	-	0.94 * (<0.001)	0.92 * (<0.001)	0.94 * (<0.001)	0.68 * (0.031)	0.85 * (0.002)	0.83 * (0.003)
CMJA	0.94 * (<0.001)	-	0.79 * (<0.001)	0.88 * (<0.001)	0.64 * (0.003)	0.92 * (<0.001)	0.90 * (<0.001)
CMJ_{RLEG}	0.92 * (<0.001)	0.79 * (<0.001)	-	0.80 * (<0.001)	0.48 * (0.033)	0.77 * (<0.001)	0.74 * (<0.001)
CMJ_{LLEG}	0.94 * (<0.001)	0.88 * (<0.001)	0.80 * (<0.001)	-	0.53 * (0.016)	0.83 * (<0.001)	0.79 * (<0.001)
SLJ	0.68 * (0.031)	0.64 * (0.003)	0.48 * (0.033)	0.53 * (0.016)	-	0.71 * (<0.001)	0.81 * (<0.001)
$PSJT_{RLEG}$	0.85 * (0.002)	0.92 * (<0.001)	0.77 * (<0.001)	0.83 * (<0.001)	0.71 * (<0.001)	-	0.93 * (<0.001)
$PSJT_{LLEG}$	0.83 * (0.003)	0.90 * (<0.001)	0.74 * (<0.001)	0.79 * (<0.001)	0.81 * (<0.001)	0.93 * (<0.001)	-

* $p < 0.05$; CMJ: bilateral countermovement jump—arms akimbo; CMJA: bilateral countermovement jump with an arm swing; CMJ_{RLEG}: unilateral countermovement jump performed with the right leg; CMJ_{LLEG}: unilateral countermovement jump performed with the left leg; SLJ: standing long jump; $PSJT_{RLEG}$: pivot step jump performed with the right leg; $PSJT_{LLEG}$: pivot step jump performed with the left leg.

4. Discussion

The purpose of this study was to investigate the reliability of the basketball-specific assessment of jumping ability, the PSJT and its validity with the jump tests commonly used to assess explosive strength and power in basketball players. Overall, the results of this study showed excellent intersession and intrasession reliability for the PSJT in young male basketball players and validity with standard jump tests in young female basketball players.

There are few studies in the literature on specific jump tests tailored to the kinematic and technical demands of a basketball game [22]. Previous studies have reported the reliability of CMJ, squat jump and drop jump tests in basketball players [25,38]. Although different methodological approaches were used, reported reliability ranged from very good to high, with CV values of 3 to 4% [25], Cronbach's alpha > 0.90 [38], and test–retest correlation of 0.98 [39]. Markovic et al. [40] also reported similar values for CMJ reliability (ICC 0.96) for one-session reliability (intrasession), while Moir et al. [41] reported equally high ICC reliability values (0.87–0.95) and CV (4.0–6.6%) for the same test between multiple sessions (intersession). The low CV values (between 1.3% and 2.4%) reported for the PSJT in the presented study suggest high reliability as well as low variation in performance between the first and subsequent trials. This also suggests an ease of administration of the test and a low motor learning effect in male basketball players.

Our finding of excellent intra- and intersession reliability was based on the mean of four consecutive trials each acquired with the mean of the two best attempts in each leg, which is consisted with previous studies [42,43]. An interesting observation is that when the test was performed with the right leg the PSJT tended to have slightly lower ICC values and higher SEM, CV, and MD values.

In the present study, all correlations between the PSJT and jump tests were strong, suggesting that the PSJT is a concurrently valid test for assessing jumping ability in basketball. This confirms past research that has demonstrated that the standing long jump and the CMJ test are the most reliable jump tests for assessing the explosive properties of the lower limbs in physically active athletes [22,40,44].

The CMJ is thought to provide an assessment of the ability to generate force rapidly during stretch-shortening cycle movements [45]. Two types are commonly used when performing the CMJ. The first involves the use of the arm swing (CMJA), while the second limits the influence of the arm swing by requiring the athlete to keep their hands on the hips [3,46]. Previous research found that the jump height is the same between CMJ and the jump shot [21]. However, in the case of PSJT, the 90-degree pivot step is followed by a fast vertical jump with a countermovement of the legs and the swinging of the arms. The apparent similarity between PSJT and CMJA is the countermovement of the legs and arms during the pivot action. When pivoting from an upright standing position, basketball players must also perform a preparatory downward movement by flexing the knees and hips before performing a rapid vertical jump. Heishman et al. [47] have shown that both CMJA and CMJ provide valid information for assessing jumping ability, but each offers distinct advantages. CMJ is useful for assessing performance changes on a long-term basis, such as changes in performance across training periods [21]. However, several authors suggest that the inclusion of an arm swing, when performing a CMJ test, leads to a higher degree of sport specificity that may improve reliability [48,49]. In the present study, the AS_{INDEX} was within the range of values reported in the past for young and adult basketball players, suggesting that the participants' intersegmental neuromuscular coordination pattern among was of a good standard [8].

From a biomechanical perspective, vertical jumps with arm swing increase jump height due to increased power and work output [8,11,50]. Specifically, the arm swing generates work in the shoulder joint and the flow of this work leads to an increase in torque in the hip joint, ultimately resulting in a higher jump height [11,12,51,52]. This mechanism may provide a basis for the finding that jump heights were higher in the PSJT tests than in the corresponding CMJ tests. In addition, previous research has shown that the effect of arm swing remains unchanged during developmental age in young basketball players [8]. Therefore, it can be hypothesized that the PSJT may also be a sport-specific jump test in adult basketball players and that future research should address this issue.

Despite the fact that the arm swing is involved in all jumping actions in basketball, previous research suggests that the CMJ had marginally larger reliability scores than the CMJA [53]. This may be due to the significant interlimb asymmetry in leg stiffness observed in basketball players performing the CMJA [54]. As leg stiffness is an important factor in generating power during jump tests [55], interlimb asymmetry in lower extremity power output results in decreased jump performance in athletes [56,57]. However, in the present study, the strong correlations when both the right and left leg were used to perform the PSJT suggest that countermovement was not affected by limb or side preference. This was also confirmed by the extracted θ_{SYM} values, where θ_{SYM} is recommended for determining differences between limbs [58]. Further confirmation of this observation was the strong and significant correlation between CMJ_{RLEG} ($r = 0.77$, $p < 0.001$) and CMJ_{LLEG} ($r = 0.83$, $p < 0.001$) with PSJT. This is consistent with the literature, as previous studies found high reliability scores for the unilateral CMJ in basketball players [59]. In addition, the performance of the PSJT with a 90° pivoting step may have favored the absence of interlimb asymmetry, as young basketball players showed a large intralimb asymmetry during a 180° change of direction test [60], with a low to moderate association with bilateral CMJ deficit in male and female players [61].

A high correlation was also found between the PSJT and the SLJ. The standing long jump is considered an indicator of maximal horizontal power generation in the sagittal plane [62]. Due to its correlation with basketball performance variables, its use to assess lower-body power in athletes is considered appropriate [44]. According to Wen et al. [62], players with high scores in SLJ are likely to be efficient in performing explosive short burst actions on the court. Pivoting with a subsequent jump shot is also a quick, powerful action that helps players gain an advantage over their opponents by rapidly moving into a position where they can score easily. Therefore, the high correlation with the SLJ likely indicates its suitability as a method for assessing sport-specific, power-related attributes in basketball.

Longitudinal examination of between-limb differences has been shown to be important for long-term monitoring of sport performance, as well as the accuracy of interpretation of asymmetry scores in bilateral and unilateral tests [63]. The lack of such monitoring is a limitation to the study. Although the literature does not support a gender bias in jump kinetics in young basketball players [64], the correlation of the PSJT with standard vertical jump tests should also be investigated and confirmed in young basketball players. In addition, the possible effect of the playing position [65] and level [66] that is evident in young basketball players was not considered. Despite the existence of contradicting findings in past research concerning the effect of the playing position on jumping performance [3], future research should consider the positional and playing level differences in the PSJT as well.

5. Conclusions

In summary, the results of this study suggest that the PSJT is a concurrent valid measure of jumping ability for estimating lower limb explosive force in young female basketball players. Its strong correlation with vertical jump tests, which are commonly used to assess jumping ability and explosive power in basketball players, makes it useful for monitoring changes in jumping performance. In addition, the movement pattern tailored to the demands of a basketball game provides coaches with an ecologically valid and reliable test for quantifying adaptation to training programs.

Author Contributions: Conceptualization, A.S.T. and H.-P.R.; methodology, A.S.T. and H.-P.R.; software, I.K.; validation, I.K. and E.B.; formal analysis, H.-P.R. and V.P.; investigation, A.S.T., H.-P.R., E.Z. and P.S.; resources, A.S.T., H.-P.R., E.Z., E.B., D.I.B. and N.K.; data curation, H.-P.R., I.K. and V.P.; writing—original draft preparation, A.S.T. and H.-P.R.; writing—review and editing, A.S.T. and V.P.; visualization, A.S.T. and V.P.; supervision, A.S.T., E.Z. and N.K.; project administration, A.S.T. and N.K. All authors have read and agreed to the published version of the manuscript.

Funding: This research received no external funding.

Institutional Review Board Statement: The study was conducted in accordance with the Declaration of Helsinki and approved by the Research Ethics and Bioethics Committee of the School of Physical Education and Sport Sciences of the National and Kapodistrian University of Athens, Greece (protocol code: 1063/13.6.2018).

Informed Consent Statement: Informed parental consent was obtained from all subjects involved in the study.

Data Availability Statement: The data presented in this study are available on reasonable request from the corresponding author.

Conflicts of Interest: The authors declare no conflict of interest.

References

1. Cabarkapa, D.; Eserhaut, D.A.; Fry, A.C.; Cabarkapa, D.V.; Philipp, N.M.; Whiting, S.M.; Downey, G.G. Relationship between Upper and Lower Body Strength and Basketball Shooting Performance. *Sports* **2022**, *10*, 139. [CrossRef] [PubMed]
2. Pehar, M.; Sekulic, D.; Sisic, N.; Spasic, M.; Uljevic, O.; Krolo, A.; Milanovic, Z.; Sattler, T. Evaluation of different jumping tests in defining position-specific and performance-level differences in high level basketball players. *Biol. Sport.* **2017**, *34*, 263–272. [CrossRef] [PubMed]
3. Ziv, G.; Lidor, R. Vertical jump in female and male basketball players: A review of observational and experimental studies. *J. Sci. Med. Sport.* **2010**, *13*, 332–339. [CrossRef] [PubMed]
4. Ghiţescu, I.G.; Tudor, V.; Moanţă, A.D. Study on the development of vertical jumping force in U18 junior basketball players. *Proc. Soc. Behav. Sci.* **2014**, *117*, 55–59. [CrossRef]
5. Morrison, M.; Martin, D.T.; Talpey, S.; Scanlan, A.T.; Delaney, J.; Halson, S.L.; Weakley, J. A systematic review on fitness testing in adult male basketball players: Tests adopted, characteristics reported and recommendations for practice. *Sport. Med.* **2022**, *52*, 1491–1532. [CrossRef]
6. Ostojic, S.M.; Mazic, S.; Dikic, N. Profiling in basketball: Physical and physiological characteristics of elite players. *J. Strength Cond. Res.* **2006**, *20*, 740–744. [CrossRef]
7. Puente, C.; Abián-Vicén, J.; Areces, F.; López, R.; Del Coso, J. Physical and physiological demands of experienced male basketball players during a competitive game. *J. Strength Cond. Res.* **2017**, *31*, 956–962. [CrossRef]
8. Gerodimos, V.; Zafeiridis, A.; Perkos, S.; Dipla, K.; Manou, V.; Kellis, S. The contribution of stretch-shortening cycle and arm-swing to vertical jumping performance in children, adolescents, and adult basketball players. *Ped. Exerc. Sci.* **2008**, *20*, 379–389. [CrossRef]
9. Harman, E.A.; Rosenstein, M.T.; Frykman, P.N.; Rosenstein, R.M.; Kraemer, W.J. Estimation of human power output from vertical jump. *J. Strength Cond. Res.* **1991**, *5*, 116–120.

10. Hoffman, J.R.; Kang, J.I.E. Evaluation of a new anaerobic power testing system. *J. Strength Cond. Res.* **2002**, *16*, 142–148.
11. Feltner, M.E.; Fraschetti, D.J.; Crisp, R.J. Upper extremity augmentation of lower extremity kinetics during countermovement vertical jumps. *J. Sport. Sci.* **1999**, *17*, 449–466. [CrossRef] [PubMed]
12. Hara, M.; Shibayama, A.; Takeshita, D.; Hay, D.C.; Fukashiro, S. A comparison of the mechanical effect of arm swing and countermovement on the lower extremities in vertical jumping. *Hum. Mov. Sci.* **2008**, *27*, 636–648. [CrossRef] [PubMed]
13. Lees, A.; Vanrenterghem, J.; De Clercq, D. Understanding how an arm swing enhances performance in the vertical jump. *J. Biomech.* **2004**, *37*, 1929–1940. [CrossRef] [PubMed]
14. Chalitsios, C.; Nikodelis, T.; Panoutsakopoulos, V.; Chassanidis, C.; Kollias, I. Classification of soccer and basketball players' jumping performance characteristics: A logistic regression approach. *Sports* **2019**, *7*, 163. [CrossRef] [PubMed]
15. Kollias, I.; Panoutsakopoulos, V.; Papaiakovou, G. Comparing jumping ability among athletes of various sports: Vertical drop jumping from 60 centimeters. *J. Strength Cond. Res.* **2004**, *18*, 546–550.
16. Laffaye, G.; Bardy, B.G.; Durey, A. Leg stiffness and expertise in men jumping. *Med. Sci. Sport. Exerc.* **2005**, *37*, 536–543. [CrossRef] [PubMed]
17. Laffaye, G.; Bardy, B.G.; Durey, A. Principal component structure and sport-specific differences in the running one-leg vertical jump. *Int. J. Sport. Med.* **2007**, *28*, 420–425. [CrossRef]
18. Panoutsakopoulos, V.; Karagianni, O.; Papaiakovou, G.; Kollias, I. Comparing drop jumping ability among elite female athletes from various sports. *J. Hum. Mov. Stud.* **2003**, *45*, 313–326.
19. Panoutsakopoulos, V.; Papachatzis, N.; Kollias, I.A. Sport specificity background affects the principal component structure of vertical squat jump performance of young adult female athletes. *J. Sport Health Sci.* **2014**, *3*, 239–247. [CrossRef]
20. Rauch, J.; Leidersdorf, E.; Reeves, T.; Borkan, L.; Elliott, M.; Ugrinowitsch, C. Different movement strategies in the countermovement jump amongst a large cohort of NBA players. *Int. J. Environ. Res. Public Health* **2020**, *17*, 6394. [CrossRef]
21. Struzik, A.; Pietraszewski, B.; Zawadzki, J. Biomechanical analysis of the jump shot in basketball. *J. Hum. Kinet.* **2014**, *42*, 73–79. [CrossRef] [PubMed]
22. Rodríguez-Rosell, D.; Mora-Custodio, R.; Franco-Márquez, F.; Yáñez-García, J.M.; González-Badillo, J.J. Traditional vs. sport-specific vertical jump tests: Reliability, validity, and relationship with the legs strength and sprint performance in adult and teen soccer and basketball players. *J. Strength Cond. Res.* **2017**, *31*, 196–206. [CrossRef] [PubMed]
23. Currell, K.; Jeukendrup, A.E. Validity, reliability and sensitivity of measures of sporting performance. *Sport. Med.* **2008**, *38*, 297–316. [CrossRef] [PubMed]
24. Gál-Pottyondy, A.; Petró, B.; Czétényi, A.; Négyesi, J.; Nagatomi, R.; Kiss, R.M. Collection and advice on basketball field tests—A literature review. *Appl. Sci.* **2021**, *11*, 8855. [CrossRef]
25. Delextrat, A.; Cohen, D. Strength, power, speed, and agility of women basketball players according to playing position. *J. Strength Cond. Res.* **2009**, *23*, 1974–1981. [CrossRef]
26. Clemente, F.M.; Praça, G.; Oliveira, R.; Aquino, R.; Araújo, R.; Silva, R.; Sarmento, H.; Afonso, J. A systematic review of the criterion validity and reliability of technical and tactical field-based tests in soccer. *Int. J. Sport. Sci. Coach.* **2022**, *17*, 1462–1487. [CrossRef]
27. Koopmann, T.; Faber, I.; Baker, J.; Schorer, J. Assessing technical skills in talented youth athletes: A systematic review. *Sport. Med.* **2020**, *50*, 1593–1611. [CrossRef]
28. Clemons, J.M.; Campbell, B.; Jeansonne, C. Validity and reliability of a new test of upper body power. *J. Strength Cond. Res.* **2010**, *24*, 1559–1565. [CrossRef]
29. Clemons, J.; Harrison, M. Validity and reliability of a new stair sprinting test of explosive power. *J. Strength Cond. Res.* **2008**, *22*, 1578–1583. [CrossRef]
30. Carter, J.E.L.; Heath, B.H. *Somatotyping: Development and Applications*; Cambridge University Press: New York, NY, USA, 1990.
31. Moore, S.A.; McKay, H.A.; Macdonald, H.; Nettlefold, L.; Baxter-Jones, A.D.; Cameron, N.; Brasher, P.M. Enhancing a somatic maturity prediction model. *Med. Sci. Sport. Exerc.* **2015**, *47*, 1755–1764. [CrossRef]
32. Linthorne, N.P. Analysis of standing vertical jumps using a force platform. *Am. J. Phys.* **2001**, *69*, 1198–1204. [CrossRef]
33. Glatthorn, J.F.; Gouge, S.; Nussbaumer, S.; Stauffacher, S.; Impellizzeri, F.M.; Maffiuletti, N.A. Validity and reliability of Optojump photoelectric cells for estimating vertical jump height. *J. Strength Cond. Res.* **2011**, *25*, 556–560. [CrossRef] [PubMed]
34. Zifchock, R.A.; Davis, I.; Higginson, J.; Royer, T. The symmetry angle: A novel, robust method of quantifying asymmetry. *Gait Posture* **2008**, *27*, 622–627. [CrossRef]
35. Wu, W.L.; Wu, J.H.; Lin, H.T.; Wang, G.J. Biomechanical analysis of the standing long jump. *Biomed. Eng.* **2003**, *15*, 186–192. [CrossRef]
36. Fleiss, J.L. *The Design and Analysis of Clinical Experiments*; John Wiley & Sons, Inc.: New York, NY, USA, 1986. [CrossRef]
37. Weir, J.P. Quantifying test-retest reliability using the intraclass correlation coefficient and the SEM. *J. Strength Cond. Res.* **2005**, *19*, 231–240. [CrossRef]
38. Erculj, F.; Blas, M.; Bracic, M. Physical demands on young elite European female basketball players with special reference to speed, agility, explosive strength, and take-off power. *J. Strength Cond. Res.* **2010**, *24*, 2970–2978. [CrossRef]
39. Duncan, M.J.; Lyons, M.; Nevill, A.M. Evaluation of peak power prediction equations in male basketball players. *J. Strength Cond. Res.* **2008**, *22*, 1379–1381. [CrossRef]

40. Markovic, G.; Dizdar, D.; Jukic, I.; Cardinale, M. Reliability and factorial validity of squat and countermovement jump tests. *J. Strength Cond. Res.* **2004**, *18*, 551–555.
41. Moir, G.; Shastri, P.; Connaboy, C. Intersession reliability of vertical jump height in women and men. *J. Strength Cond. Res.* **2008**, *22*, 1779–1784. [CrossRef]
42. Doyle, T.L.; Newton, R.U.; Burnett, A.F. Reliability of traditional and fractal dimension measures of quiet stance center of pressure in young, healthy people. *Arch. Phys. Med. Rehabil.* **2005**, *86*, 2034–2040. [CrossRef]
43. Pinsault, N.; Vuillerme, N. Test–retest reliability of centre of foot pressure measures to assess postural control during unperturbed stance. *Med. Eng. Phys.* **2009**, *31*, 276–286. [CrossRef] [PubMed]
44. McCormick, B.T.; Hannon, J.C.; Newton, M.; Shultz, B.; Detling, N.; Young, W.B. The effects of frontal-and sagittal-plane plyometrics on change-of-direction speed and power in adolescent female basketball players. *Int. J. Sport. Physiol. Perform.* **2016**, *11*, 102–107. [CrossRef] [PubMed]
45. Komi, P.V. Stretch-shortening cycle: A powerful model to study normal and fatigued muscle. *J. Biomech.* **2000**, *33*, 1197–1206. [CrossRef] [PubMed]
46. Watkins, C.M.; Barillas, S.R.; Wong, M.A.; Archer, D.C.; Dobbs, I.J.; Lockie, R.G.; Coburn, J.W.; Tran, T.T.; Brown, L.E. Determination of vertical jump as a measure of neuromuscular readiness and fatigue. *J. Strength Cond. Res.* **2017**, *31*, 3305–3310. [CrossRef]
47. Heishman, A.D.; Daub, B.D.; Miller, R.M.; Freitas, E.D.; Frantz, B.A.; Bemben, M.G. Countermovement jump reliability performed with and without an arm swing in NCAA division 1 intercollegiate basketball players. *J. Strength Cond. Res.* **2020**, *34*, 546–558. [CrossRef] [PubMed]
48. Slinde, F.; Suber, C.; Suber, L.; Edwén, C.E.; Svantesson, U. Test-retest reliability of three different countermovement jumping tests. *J. Strength Cond. Res.* **2008**, *22*, 640–644. [CrossRef]
49. Vaverka, F.; Jandačka, D.; Zahradník, D.; Uchytil, J.; Farana, R.; Supej, M.; Vodičar, J. Effect of an arm swing on countermovement vertical jump performance in elite volleyball players. *J. Hum. Kinet.* **2016**, *53*, 41–50. [CrossRef]
50. Walsh, M.S.; Böhm, H.; Butterfield, M.M.; Santhosam, J. Gender bias in the effects of arms and countermovement on jumping performance. *J. Strength Cond. Res.* **2007**, *21*, 362–366. [CrossRef]
51. Domire, Z.J.; Challis, J.H. An induced energy analysis to determine the mechanism for performance enhancement as a result of arm swing during jumping. *Sport. Biomech.* **2010**, *9*, 38–46. [CrossRef]
52. Giatsis, G.; Panoutsakopoulos, V.; Kollias, I.A. Biomechanical differences of arm swing countermovement jumps on sand and rigid surface performed by elite beach volleyball players. *J. Sport. Sci.* **2018**, *36*, 997–1008. [CrossRef]
53. Heishman, A.; Daub, B.; Miller, R.; Brown, B.; Freitas, E.; Bemben, M. Countermovement jump inter-limb asymmetries in collegiate basketball players. *Sports* **2019**, *7*, 103. [CrossRef] [PubMed]
54. Struzik, A.; Winiarski, S.; Zawadzki, J. Inter-limb asymmetry of leg stiffness in national second-league basketball players during countermovement jumps. *Symmetry* **2022**, *14*, 440. [CrossRef]
55. Korff, T.; Horne, S.L.; Cullen, S.J.; Blazevich, A.J. Development of lower limb stiffness and its contribution to maximum vertical jumping power during adolescence. *J. Exp. Biol.* **2009**, *212*, 3737–3742. [CrossRef] [PubMed]
56. Bailey, C.; Sato, K.; Alexander, R.; Chiang, C.Y.; Stone, M.H. Isometric force production symmetry and jumping performance in collegiate athletes. *J. Trainology* **2013**, *2*, 1–5. [CrossRef] [PubMed]
57. Bell, D.R.; Sanfilippo, J.L.; Binkley, N.; Heiderscheit, B.C. Lean mass asymmetry influences force and power asymmetry during jumping in collegiate athletes. *J. Strength Cond. Res.* **2014**, *28*, 884–891. [CrossRef] [PubMed]
58. Bishop, C.; Read, P.; Chavda, S.; Turner, A. Asymmetries of the lower limb: The calculation conundrum in strength training and conditioning. *Strength Cond. J.* **2016**, *38*, 27–32. [CrossRef]
59. Pérez-Castilla, A.; García-Ramos, A.; Janicijevic, D.; Delgado-García, G.; De la Cruz, J.C.; Rojas, F.J.; Cepero, M. Between-session reliability of performance and asymmetry variables obtained during unilateral and bilateral countermovement jumps in basketball players. *PLoS ONE* **2021**, *16*, e0255458. [CrossRef]
60. Šarabon, N.; Smajla, D.; Maffiuletti, N.A.; Bishop, C. Strength, jumping and change of direction speed asymmetries in soccer, basketball and tennis players. *Symmetry* **2020**, *12*, 1664. [CrossRef]
61. Kozinc, Ž.; Šarabon, N. Bilateral deficit in countermovement jump and its association with change of direction performance in basketball and tennis players. *Sport. Biomech.* 2021, 1–14, *in press*. [CrossRef]
62. Wen, N.; Dalbo, V.J.; Burgos, B.; Pyne, D.B.; Scanlan, A.T. Power testing in basketball: Current practice and future recommendations. *J. Strength Cond. Res.* **2018**, *32*, 2677–2691. [CrossRef]
63. Bishop, C.; Read, P.; Lake, J.; Chavda, S.; Turner, A. Interlimb asymmetries: Understanding how to calculate differences from bilateral and unilateral tests. *Strength Cond. J.* **2018**, *40*, 1–6. [CrossRef]
64. Walsh, M.S.; Waters, J.A.; Böhm, H.; Potteiger, J.A. Gender bias in jumping kinetics in national collegiate athletic association division I basketball players. *J. Strength Cond. Res.* **2007**, *21*, 958–962. [CrossRef] [PubMed]
65. te Wierike, S.C.M.; Elferink-Gemser, M.T.; Tromp, E.J.Y.; Vaeyens, R.; Visscher, C. Role of maturity timing in selection procedures and in the specialisation of playing positions in youth basketball. *J. Sport. Sci.* **2015**, *33*, 337–345. [CrossRef] [PubMed]
66. Torres-Unda, J.; Zarrazquin, I.; Gil, J.; Ruiz, F.; Irazusta, A.; Kortajarena, M.; Seco, J.; Irazusta, J. Anthropometric, physiological and maturational characteristics in selected elite and non-elite male adolescent basketball players. *J. Sport. Sci.* **2013**, *31*, 196–203. [CrossRef]

Journal of
Functional Morphology and Kinesiology

MDPI

Article

Congested Period in Professional Youth Soccer Players Showed a Different High Decelerations Profile in the Group Performance and a Specific Positional Behaviour

Borja Muñoz-Castellanos [1], Alberto Rabano-Muñoz [1], Bernardo Requena [2], Luis Suarez-Arrones [2,3,4] and Jose A. Asian-Clemente [1,2,3,*]

1 Real Betis Balompié, Performance Department, 41012 Seville, Spain
2 Football Science Institute, 18016 Granada, Spain
3 Department of Sport Sciences, Universidad Pablo de Olavide, 41013 Seville, Spain
4 FC Lugano, Performance Departament, 6900 Lugano, Switzerland
* Correspondence: jasicle@upo.es; Tel.: +34-954977601

Abstract: Present soccer demands are increasing in terms of running requirements and the number of matches until youth soccer players experience several periods of fixture congestion during the season. Currently, congested periods have not been extensively studied in this population. For this reason, this study aimed to compare the running demands of professional youth soccer players in congested periods according to their specific playing positions. Twenty youth players were grouped according to their position: Central Defenders (CD), Fullbacks (FB), Midfielders (MF), Wide Midfielders (WM) and Strikers (ST). A GPS system was used to monitor the players during the first (M1), second (M2) and third (M3) matches played during a congested period, measuring their total distance covered (TDC), DC 18.0–20.9 $km \cdot h^{-1}$, DC 21.0–23.9 $km \cdot h^{-1}$, DC > 24.0 $km \cdot h^{-1}$, number of high accelerations (>2.5 $m \cdot s^{-2}$), number of high decelerations (<2.5 $m \cdot s^{-2}$) and peak speed ($km \cdot h^{-1}$). M1, M2 and M3 showed the same TDC, DC 18.0–20.9 $km \cdot h^{-1}$, DC 21.0–23.9 $km \cdot h^{-1}$, DC > 24.0 $km \cdot h^{-1}$, number of high accelerations, and peak speed ($p > 0.05$). The statistical analysis showed significant differences between M1, M2 and M3 in the decelerations recorded between M1 and M3 ($p < 0.05$). Likewise, each position showed specific behaviours during the congested period, with all showing at least one difference in DC 18.0–20.9 $km \cdot h^{-1}$, 21.0–23.9 $km \cdot h^{-1}$ or >24.0 $km \cdot h^{-1}$ between M1, M2 and M3 ($p < 0.05$). In conclusion, coaches should pay attention to the fatigue produced by the number of high decelerations. Secondly, an individualized training protocol should be considered according to the running requirements of each position when youth professional soccer players are involved in a congested period.

Keywords: time-motion; running; fatigue; external load; GPS

Citation: Muñoz-Castellanos, B.; Rabano-Muñoz, A.; Requena, B.; Suarez-Arrones, L.; Asian-Clemente, J.A. Congested Period in Professional Youth Soccer Players Showed a Different High Decelerations Profile in the Group Performance and a Specific Positional Behaviour. *J. Funct. Morphol. Kinesiol.* **2022**, *7*, 108. https://doi.org/10.3390/jfmk7040108

Academic Editor: Diego Minciacchi

Received: 27 October 2022
Accepted: 25 November 2022
Published: 29 November 2022

1. Introduction

Soccer is a highly intermittent sport where players' activity is composed of high and low-intensity movements of varying lengths and situations depending upon an array of factors [1–4]. One of the most important aspects of modern elite soccer is the increasing demands in terms of running requirements and the number of matches played during the season [5]. The number of matches has increased in top-level European teams from around 50 matches in the 2008/2009 season to around 60 matches currently [6], meaning that professional players experience several periods of fixture congestion during the season [7]. A congested schedule is considered to exist when there is a minimum of two successive bouts of match-play, with an inter-match recovery period of less than 96 h [8]. Congested periods are common when soccer teams participate in more than one competition during the season or when national teams play in an international tournament [9–11].

Previous studies have shown that depending on the variable examined, professional soccer players may experience changes in their running requirements between matches during congested and non-congested periods. During congested periods, it has been affirmed that the distances covered at different running intensities in elite professional soccer remained unchanged between matches [12–16]. However, when congested periods are experienced, some researchers have found a difference in performance in terms of total distance, number of sprints, distance covered at different intensities, and acceleration and deceleration profiles [17–19]. Although congested periods are a common concern in elite professional soccer [20], they have not been extensively studied in the younger soccer population. In professional soccer academies, this type of schedule has recently received attention in specific age groups: Under 14, Under 15, Under 17 and Under 19 [21–24]. In these age groups, there were no differences between congested match periods and non-congested periods for the total distance and the distance covered >21 km·h^{-1} in the most demanding passages of the matches [23]. Variables including the numbers of accelerations and decelerations and mean metabolic power showed increases in congested periods compared to non-congested periods [22]. Another study indicated a decline in total distance covered and player load during the congested period [24]. Although these previous studies compared the differences between congested and non-congested periods, other investigations have analysed player behaviour during matches in the same congested periods. Some researchers found dissimilarities in the distance covered at different intensities, but no changes were found in sprint frequency [25], and the number of decelerations decreased in the last four matches played in a row in a congested period by elite youth soccer players [22]. In contrast, other studies showed that running performance at different intensities was maintained between matches 1 and 3 for under-23 soccer players [26]. Studies on this topic showed discrepancies in the results obtained. As some authors have found, these controversial outcomes may be due to a high variability between soccer matches [27–29]. This diversity in running performances between matches has been proven to be relatively "large" [30], reaching from 15% to >60% depending on the variable studied [31], thus highlighting the inherent context of soccer.

Similar congested periods occur when youth soccer players are evaluated in tournaments carried out over a short period of 3–4 days [22,32]. In a short tournament of this kind, there were no alterations found in the total distance covered, high-intensity running distance, or maximum running speed [8,22,33,34]. However, another study showed that accelerations were affected in this kind of short tournament [32].

Taking into account that the external load of soccer players is related to their playing position [27,28,35,36], the scientific literature has also evaluated running behaviour for different playing positions during congested periods. These investigations showed that there were differences between specific positions, especially in the total distance covered and the distance covered at moderate and high intensity [15,18,19,26,37].

To the authors' knowledge, there is no research to date that has investigated the running performance of different playing positions for youth soccer players during a congested period. These periods are common in professional soccer, and they have received much attention in recent times due to an increase in matches in both national and international championships. In youth soccer players, these congested periods have not been studied extensively during a regular championship. This may be because these types of periods have not previously been experienced by young players. According to the rising interest in performance development, youth soccer players have been involved in systematic training programs and matches designed to progress player promotion [38]. In order to better understand the physical demands required to play professional soccer, it is necessary to delineate the match demands in this framework [39]. These types of investigations are needed to clarify running requirements in youth players during adult soccer situations. The aim of this study was thus to compare the external load of professional youth soccer players in matches played during a congested period and evaluate the positional running requirements for each match.

2. Materials and Methods

2.1. Subjects

Twenty male youth soccer players from one Spanish professional academy participated in this study (Age = 15.95 ± 1.85 years, Height = 175.6 ± 5.35 cm, Body Weight = 63.17 ± 6.9 kg). A total of 4 players were selected from the Under-14 team (mean ± SD, 13.25 ± 0.5 years; 172.53 ± 4.12 cm; 56.95 ± 4.29 kg), 7 from the Under-16 team (mean ± SD, 15.29 ± 0.49 years; 172.93 ± 4.55 cm; 59.01 ± 5.43 kg) and 9 from the Under-19 team (mean ± SD, 17.67 ± 0.71 years; 179.04 ± 4.66 cm; 69.18 ± 3.08 kg). They were grouped according to their positional roles as Central Defenders (CD, n = 6), Fullbacks (FB, n = 5), Midfielders (MF, n = 3), Wide Midfielders (WM, n = 3) and Strikers (ST, n = 3). All players, regardless of their team or position, participated in 5 soccer training sessions per week (strength and conditioning and technical-tactical sessions) and normally competed in a single weekly match except during the congested period where they played three matches in the same week. All players were declared injury-free and fit for competition by medical staff prior to participation in any match. These data were acquired as a condition of player monitoring, in which player activities are measured over the competitive season [40], so ethics committee clearance was not required. Nevertheless, the study conformed to the recommendations of the Declaration of Helsinki, and the participants were informed of the study's design and aims, giving their consent before it started.

2.2. Design

The running demands of 21 official matches (Under 14 = 3, Under 16 = 9 and Under 19 = 9) during the 2020–2021 season were monitored, resulting in 60 player match observations. These observations were subsequently partitioned into 3 different types of matches according to their temporal distribution and were categorized as M1 for the first matches of congested periods, M2 for the second and M3 for the third. Only 7-day microcycles were included in this study, excluding those where the matches were disputed with a duration of more than 72 h with each other. Players were only included in the analysis if they played for ≥75% of the total match duration [22,32]. Those who did not fulfil this criterion were excluded from the data analysis. All assessed matches were part of the Regular Championship for each team, and there were different match durations for each age category: 2 halves of 40 min with a 15-min half-time interval for those Under 14, and 2 halves of 45 min with a 15-min half-time interval for those Under 16 and those Under 19. All matches were played on artificial grass and under similar environmental conditions. Substitutions were also different for each age category: for the Under-14 group, rotative substitutions were permitted, while coaches allowed the Under-16 and Under-19 groups a maximum of 5 substitutions, respectively. None of the teams used systematic post-match recovery regimens between the assessed matches.

External load was monitored using global positioning system (GPS) devices (10-Hz, Catapult Sports, Melbourne, Australia). The devices were fitted to the upper back of each player using an elastic harness (Catapult Sports, Melbourne, Australia). The reliability and accuracy of 10 Hz GPS devices have been reported previously [41]. The studied variables comprise the following: total distance covered (TDC), distance covered while running at high speeds (DC 18.0–20.9 $km \cdot h^{-1}$), distance covered while running at very high speeds (DC 21.0–23.9 $km \cdot h^{-1}$), distance covered while sprinting (DC > 24.0 $km \cdot h^{-1}$), numbers of high accelerations (>2.5 $m \cdot s^{-2}$), numbers of high decelerations (<2.5 $m \cdot s^{-2}$) and peak speed ($km \cdot h^{-1}$). All the variables analysed were expressed in relative values per minute. Variables were classified in accordance with previous studies [42,43].

2.3. Statistics

Data are presented as mean ± standard deviation (SD). All variables presented a normal distribution (Shapiro-Wilk Test). A one-way analysis of variance (ANOVA) was used to determine differences between teams and playing positions. In the event of a significant difference, Bonferroni's post hoc tests were used to identify any localized effects.

Differences between groups and positions were analysed for practical significance using magnitude-based inferences by pre-specifying a 0.2 between-subject SD as the smallest worthwhile effect [44]. The standardized difference or effect size (ES, 90% confidence limit [90%CL]) in the selected variables was calculated. Threshold values for assessing the magnitudes of the ES (changes as a fraction or multiple of baseline standard deviation) were <0.20, 0.20, 0.60, 1.2 and 2.0 for trivial, small, moderate, large and very large, respectively [44].

3. Results

Comparisons of the external loads during the three matches in congested fixtures are shown in Table 1 and Figure 1. The statistical analysis showed significant differences in the numbers of high decelerations between M1 and M3 ($p < 0.05$).

Table 1. External load during the three matches in a congested fixture.

	Values		*p* Value	%	Q
M1 vs. M2					
TDC ($m \cdot min^{-1}$)	104.4 ± 8.1	106.1 ± 7.9	0.28	58/39/3	Possibly, may (not)
DC 18.0–20.9 $km \cdot h^{-1}$ ($m \cdot min^{-1}$)	4.6 ± 1.5	4.5 ± 1.6	0.36	44/53/3	Possibly, may (not)
DC 21.0–23.9 $km \cdot h^{-1}$ ($m \cdot min^{-1}$)	2.6 ± 1.0	2.5 ± 1.2	0.27	61/35/3	Possibly, may (not)
DC > 24.0 $km \cdot h^{-1}$ ($m \cdot min^{-1}$)	1.4 ± 0.6	1.3 ± 0.9	0.30	70/23/7	Possibly, may (not)
Acc > 2.5 $m \cdot s^{-2}$ ($n \cdot min^{-1}$)	0.10 ± 0.05	0.12 ± 0.07	0.26	36/63/1	Possibly, may (not)
Dec > 2.5 $m \cdot s^{-2}$ ($n \cdot min^{-1}$)	0.05 ± 0.03	0.06 ± 0.05	0.14	68/31/1	Possibly, may (not)
Peak speed ($km \cdot h^{-1}$)	29.4 ± 1.5	28.6 ± 2.2	0.08	87/12/1	Likely, probable
M1 vs. M3					
TDC ($m \cdot min^{-1}$)	104.4 ± 8.1	106.7 ± 12.3	0.34	61/34/6	Possibly, may (not)
DC 18.0–20.9 $km \cdot h^{-1}$ ($m \cdot min^{-1}$)	4.6 ± 1.5	4.3 ± 1.7	0.14	72/27/1	Possibly, may (not)
DC 21.0–23.9 $km \cdot h^{-1}$ ($m \cdot min^{-1}$)	2.6 ± 1.0	2.6 ± 0.9	0.69	26/66/8	Possibly, may (not)
DC > 24.0 $km \cdot h^{-1}$ ($m \cdot min^{-1}$)	1.4 ± 0.6	1.5 ± 1.0	0.50	54/35/11	Possibly, may (not)
Acc > 2.5 $m \cdot s^{-2}$ ($n \cdot min^{-1}$)	0.10 ± 0.05	0.10 ± 0.06	0.99	13/74/13	Unlikely, probable
Dec > 2.5 $m \cdot s^{-2}$ ($n \cdot min^{-1}$)	0.05 ± 0.03	0.07 ± 0.04	0.04	82/18/0	Likely, probable
Peak speed ($km \cdot h^{-1}$)	29.4 ± 1.5	28.7 ± 2.0	0.62	88/12/0	Likely, probable
M2 vs. M3					
TDC ($m \cdot min^{-1}$)	106.1 ± 7.9	106.7 ± 12.3	0.88	27/55/19	Possibly, may (not)
DC 18.0–20.9 $km \cdot h^{-1}$ ($m \cdot min^{-1}$)	4.5 ± 1.6	4.3 ± 1.7	0.53	44/48/8	Possibly, may (not)
DC 21.0–23.9 $km \cdot h^{-1}$ ($m \cdot min^{-1}$)	2.5 ± 1.2	2.6 ± 0.9	0.42	49/46/5	Possibly, may (not)
DC > 24.0 $km \cdot h^{-1}$ ($m \cdot min^{-1}$)	1.3 ± 0.9	1.5 ± 1.0	0.41	58/11/31	Possibly, may (not)
Acc > 2.5 $m \cdot s^{-2}$ ($n \cdot min^{-1}$)	0.12 ± 0.07	0.10 ± 0.06	0.39	3/57/40	Possibly, may (not)
Dec > 2.5 $m \cdot s^{-2}$ ($n \cdot min^{-1}$)	0.06 ± 0.05	0.07 ± 0.04	0.17	8/92/0	Unlikely, probable
Peak speed ($km \cdot h^{-1}$)	28.6 ± 2.2	28.7 ± 2.0	0.81	14/79/6	Possibly, may (not)

Data are presented as mean ± SD; % = percentage of change; Q = qualitative value; TDC = Total distance covered; Acc = Accelerations; Dec = Decelerations.

The running activity of each position is shown in Table 2, Figures 2 and 3. The analysis indicated that in at least one variable of DC 18.0–20.9 $km \cdot h^{-1}$, 21.0–23.9 $km \cdot h^{-1}$ and >24.0 $km \cdot h^{-1}$ all positions showed significant differences in the three matches during the congested period ($p < 0.05$). The MF achieved a significantly higher DC 18.0–20.9 $km \cdot h^{-1}$ in M3 than in the other matches (MF:M1 vs. M3, $p = 0.04$, ES = 0.12; MF:M2 vs. M3, $p = 0.02$, ES = 0.16), and the ST achieved a significantly higher TDC in M2 than in M3 (ST:M2 vs. M3, $p = 0.03$, ES = −0.39). The FB and CD achieved a significantly higher DC 21.0–23.9 $km \cdot h^{-1}$ in M1 than M3 and M2, respectively (FB:M1 vs. M3, $p = 0.03$, ES = −0.42; CD:M1 vs. M2, $p = 0.01$, ES = −0.52). Similarly, the MF achieved a significantly higher DC 21.0–23.9 $km \cdot h^{-1}$ in M3 than in M1 (MF:M1 vs. M3, $p = 0.00$, ES = 0.23). In M3, a significantly lower DC > 24.0 $km \cdot h^{-1}$ was found than in M1 for the CD (CD:M1 vs. M3, $p = 0.01$, ES = −0.35) and ST (ST:M1 vs. M3, $p = 0.00$, ES = −0.08) positions, but these were higher than in M2 for the MF (MF:M2 vs. M3, $p = 0.00$, ES = 0.23). The WM showed significantly more DC > 24.0 $km \cdot h^{-1}$ in M2 and M3 than in M1 (WM:M1 vs. M2, $p = 0.00$, ES = 0.89; WM:M1 vs. M3, $p = 0.00$, ES:1.03p). The WM and MF exhibited significantly more high decelerations in M3 than in M1 (WM:M1 vs. M3, 0.00, ES = 01.85; MF:M1 vs. M3, $p = 0.00$, ES = 0.56). The CD (CD:M2 vs. M3, $p = 0.03$, ES = 0.07), ST (ST:M2 vs. M3, $p = 0.03$, ES = 0.47) and MF (MF:M2 vs. M3, $p = 0.00$, ES = 0.35) also showed more high decelerations in M3 than in M2.

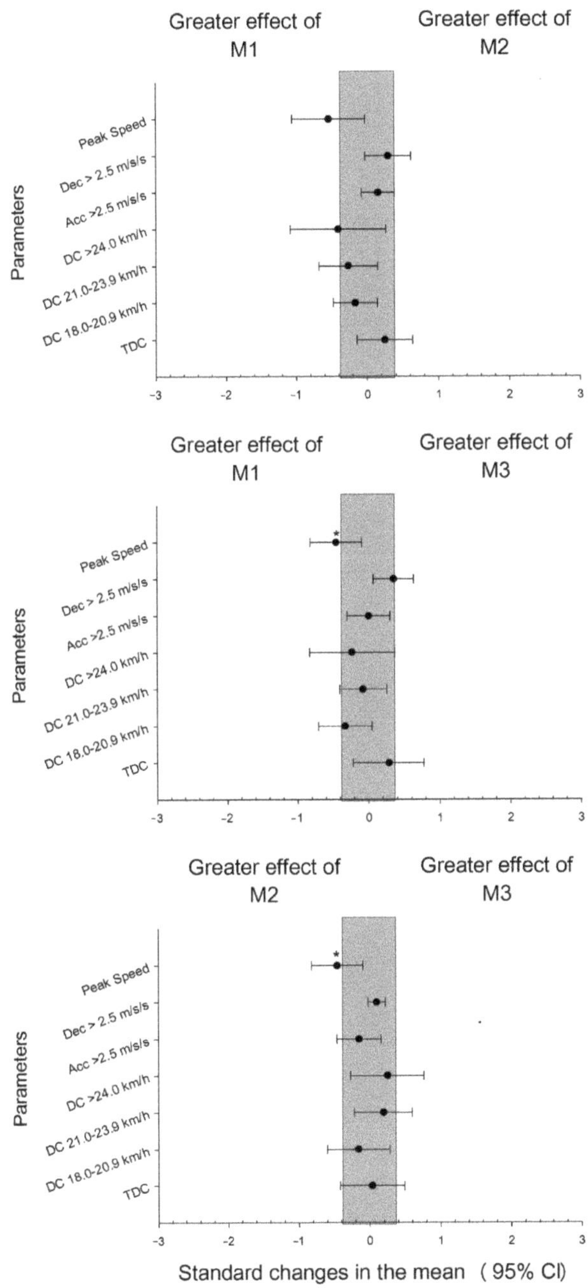

Figure 1. Comparison of external load during three matches in a congested fixture. * $p < 0.05$.

Table 2. External load during the three matches in a congested fixture according to the playing position.

		TDC ($m \cdot min^{-1}$)	DC 18.0–20.9 $km \cdot h^{-1}$ ($m \cdot min^{-1}$)	DC 21.0–23.9 $km \cdot h^{-1}$ ($m \cdot min^{-1}$)	DC > 24.0 $km \cdot h^{-1}$ ($m \cdot min^{-1}$)	Acc > 2.5 $m \cdot s^{-2}$ ($n \cdot min^{-1}$)	Dec > 2.5 $m \cdot s^{-2}$ ($n \cdot min^{-1}$)	Peak Speed ($km \cdot h^{-1}$)
CD	M1	103.5 ± 8.4	3.7 ± 0.2	2.2 ± 0.6	1.2 ± 0.7	0.10 ± 0.04	0.04 ± 0.03	29.1 ± 1.9
	M2	103.0 ± 4.1	3.1 ± 0.8	1.4 ± 0.5	0.9 ± 0.1	0.11 ± 0.05	0.03 ± 0.01	29.1 ± 1.6
	M3	111.9 ± 14.4	3.8 ± 2.0	2.0 ± 1.0	0.8 ± 0.7	0.06 ± 0.03	0.04 ± 0.01	27.0 ± 1.6
FB	M1	99.9 ± 7.2	5.0 ± 1.5	2.9 ± 0.4	1.3 ± 0.4	0.09 ± 0.05	0.05 ± 0.03	29.7 ± 1.1
	M2	106.8 ± 4.7	4.7 ± 0.9	2.7 ± 0.7	1.3 ± 0.7	0.12 ± 0.08	0.07 ± 0.04	29.6 ± 1.3
	M3	106.4 ± 3.0	4.2 ± 0.5	2.3 ± 0.5	1.5 ± 0.5	0.09 ± 0.01	0.06 ± 0.03	29.5 ± 1.0
MF	M1	106.3 ± 3.8	3.5 ± 1.1	1.4 ± 0.3	0.9 ± 0.3	0.07 ± 0.05	0.03 ± 0.01	28.6 ± 0.6
	M2	110.8 ± 11.0	3.6 ± 1.8	2.1 ± 1.1	0.8 ± 0.5	0.10 ± 0.08	0.04 ± 0.02	27.3 ± 1.7
	M3	116.0 ± 16.5	4.8 ± 2.1	2.5 ± 0.9	1.2 ± 0.9	0.07 ± 0.03	0.06 ± 0.03	27.8 ± 1.9
WM	M1	103.9 ± 11.0	5.2 ± 0.8	3.5 ± 0.3	1.1 ± 0.4	0.15 ± 0.01	0.03 ± 0.01	28.9 ± 2.1
	M2	104.6 ± 13.5	5.9 ± 1.6	4.1 ± 1.6	2.7 ± 0.4	0.19 ± 0.07	0.16 ± 0.04	29.9 ± 0.4
	M3	99.9 ± 12.0	5.2 ± 1.4	3.8 ± 0.4	3.2 ± 0.8	0.22 ± 0.06	0.13 ± 0.01	29.3 ± 1.9
ST	M1	97.1 ± 7.8	4.3 ± 1.0	2.7 ± 0.6	2.0 ± 0.2	0.10 ± 0.04	0.10 ± 0.01	31.1 ± 2.1
	M2	101.8 ± 5.7	4.5 ± 0.1	2.8 ± 0.4	1.6 ± 1.3	0.10 ± 0.05	0.09 ± 0.01	28.7 ± 3.1
	M3	91.4 ± 3.5	3.1 ± 1.0	2.3 ± 0.5	1.6 ± 0.1	0.09 ± 0.02	0.12 ± 0.02	30.7 ± 3.0

Data are presented as mean ± SD. TDC = Total distance covered; Acc = Accelerations; Dec = Decelerations; CD = Central Defenders; FB = Fullbacks; MF = Midfielders; WM = Wide Midfielders; ST = Strikers.

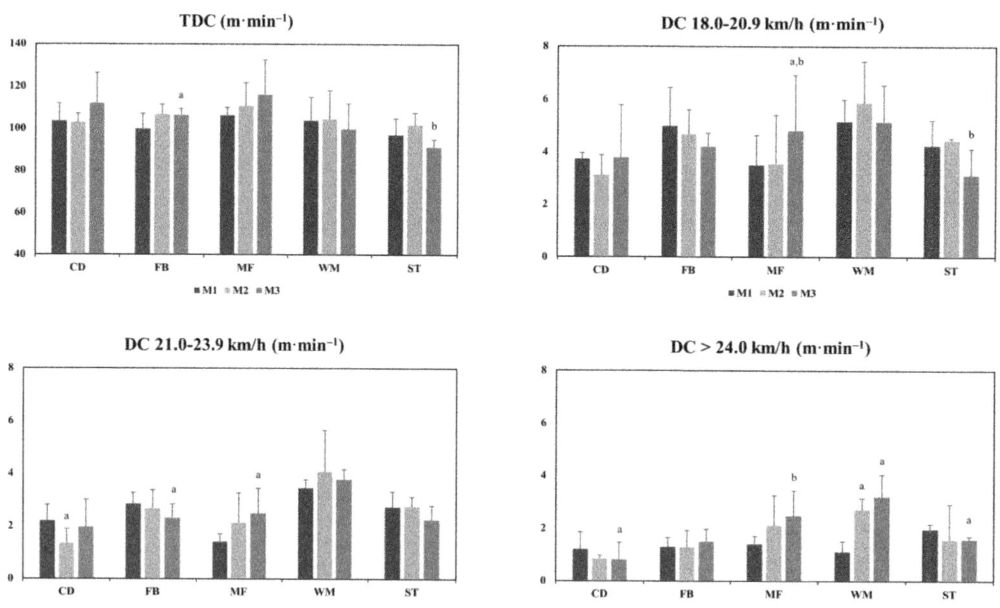

Figure 2. Comparison of distances covered at different speed thresholds during the three matches in a congested fixture. TDC = Total distance covered; CD = Central Defenders; FB = Fullbacks; MF = Midfielders; WM = Wide Midfielders; ST = Strikers; a = significant difference with M1; b = significant difference with M2.

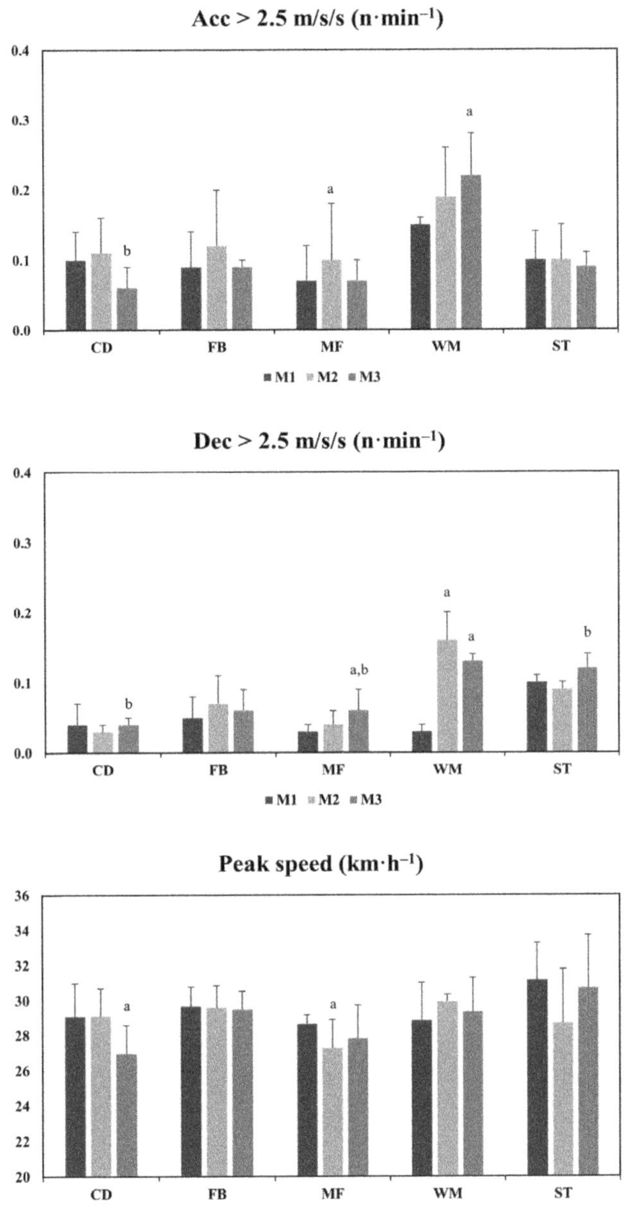

Figure 3. Comparison of accelerations, decelerations and peak speed during the three matches in a congested fixture. Acc = Accelerations; Dec = Decelerations; CD = Central Defenders; FB = Fullbacks; MF = Midfielders; WM = Wide Midfielders; ST = Strikers; a = significant difference with M1; b = significant difference with M2.

4. Discussion

This study aimed to compare the external load of professional youth soccer players in matches played in a congested period and evaluate the positional running requirements of each match. This is the first research undertaken during the Regular Championship to assess the running performance of Under 14, Under 16 and Under 19 elite soccer players in a congested schedule. Our results showed that when playing three matches in the same week, youth soccer players did not show differences in their running performance except in the number of decelerations between M1 and M3. There were differences in the running requirements based on their playing position, showing important variability in their behaviour during a congested period.

The soccer players involved in our intervention did not display differences in total TDC, DC 18–20.9 km·h^{-1}, DC 21.0–23.9 kh·h^{-1}, DC > 24.0 km·h^{-1}, numbers of high accelerations >2.5 m·s^{-2} and peak speed between the matches in a congested calendar. An important issue that could explain the similarity of our studied variables is that when soccer players have to participate in three matches during the same week, they may consciously adopt a pacing strategy to maintain high-intensity actions [8,45], achieving relatively uniform performances between matches. Although most of the studied variables showed no significant change, differences in the number of high decelerations between M1 and M3 were found. High decelerations have been highlighted as an important variable for understanding the physical demands on team sport players [46], but currently, despite the great interest in deceleration demands in soccer, there is a lack of a practical and concise approach to scientific research [47]. Taking into account that deceleration loads provoke high mechanical and metabolic demands [48,49] and are related to post-match muscle damage and fatigue [50], soccer coaches should pay attention to youth players' recovery during a congested period, especially as the number of matches increases. Unfortunately, the contextual variables of the analysed matches were not studied. Considering that context can influence match outcomes [51–53], these aspects may have influenced the players' behaviour.

It is difficult to compare the results obtained in this study with the previously published literature because there is a controversial and limited amount of available research in this area. While some authors have affirmed that soccer players showed increased running performances between the matches of a congested period [25], others have indicated that running activity was reduced [18]. Differences between our data and the previous studies could be because the sample in our study comprised Under 14, Under 16 and Under 19 professional youth players, while other studies monitored professional adult soccer players. The literature has demonstrated that youth and professional soccer players have different running behaviours when their match performances are assessed, affirming that data derived from a given population may not be relevant to other populations [54].

This investigation emphasized the positional running requirements developed in a congested period, revealing differences in all positions between the matches for some of the studied variables. Our data are in line with the published literature, finding that all positions showed differences in some parameters related to high-speed running (DC 18.0–20.9 km·h^{-1}, 21.0–23.9 km·h^{-1} or >24.0 km·h^{-1}). These findings are also in line with a recent investigation that analysed the inter-position diversity between match changes in measures of match physical performance, which demonstrated a higher match-to-match variation in distance covered using high-intensity running (≥18 km·h^{-1}) [29,31]. Similar findings have been found in highly trained youth soccer players, where the between-match variability in high-intensity and very high-intensity activities was substantially higher than for the total distance covered [30]. A recent systematic review of match demands in youth soccer has demonstrated that running requirements demands were specific to the player's position [54]. In the same way, our results reinforce the idea of specific positional running requirements during congested periods [55–57]. These results highlight the importance of contextualizing matches and attending to variables such as formations, score lines or home-away locations because they affect running performance [8].

Although the current investigation adds insightful information about the external load during a congested period in soccer, some limitations must be considered. A small sample of players was included in the analysis. The need to be on the field for at least 75% of the total time in three matches together with the use of players from a professional academy limited the accuracy of the sample. Secondly, this study has been carried out with fixed speed thresholds. Even though these thresholds provide useful information regarding player development and allow direct outcome comparisons between studies [54], further investigations should check what happens when individualizing speed thresholds are used to provide a more accurate representation of match running loads in youth soccer players. Another important point to note is that contextual factors such as the standard of the opponents, scores, period during the season and classification were not considered. Given that these aspects influence the running behaviour of soccer players, future studies should be performed, adding these elements. Lastly, this study did not measure any internal load variables that could minimize player fatigue. In physiological terms, the possible effects of genetics on athletic performance could be taken into account [58]. Future studies should try to evaluate the internal load and effects of genetics by increasing the sample during congested periods.

5. Conclusions

This study found differences only in the number of high decelerations in matches in congested periods. Coaches should therefore pay special attention to the fatigue produced by this variable. Further investigations should further assess whether decelerations are decisive for the overall performance of the team during periods of maximum intensity such as congested periods. Secondly, significant differences in running requirements have been found between matches according to player position. Consequently, individualized training protocols should be used when professional youth soccer players experience a congested period. Thus, the scientific evidence revealed by this investigation responds to the complexity of interactions established between extrinsic and intrinsic factors relating to players and teams, thereby providing tools to ensure efficient training through specific stimuli [59].

Author Contributions: Conceptualization, B.M.-C. and J.A.A.-C.; methodology, B.M.-C.; software, A.R.-M.; validation, B.M.-C., J.A.A.-C. and A.R.-M.; formal analysis, J.A.A.-C.; investigation, J.A.A.-C.; resources, B.M.-C.; data curation, A.R.-M.; writing—original draft preparation, B.M.-C.; writing—review and editing, J.A.A.-C.; visualization, B.R.; supervision, L.S.-A. All authors have read and agreed to the published version of the manuscript.

Funding: This research received no external funding.

Institutional Review Board Statement: These data were acquired as a condition of player monitoring, in which player activities are measured over the competitive season [35], so ethics committee clearance was not required. Nevertheless, the study conformed to the recommendations of the Declaration of Helsinki, and the participants were informed of the study's design and aims, giving their consent before it started.

Informed Consent Statement: Informed consent was obtained from all subjects involved in the study.

Data Availability Statement: Not applicable.

Conflicts of Interest: The authors declare no conflict of interest.

References

1. Orendurff, M.S.; Walker, J.D.; Jovanovic, M.; Tulchin, K.L.; Levy, M.; Hoffmann, D.K. Intensity and duration of intermittent exercise and recovery during a soccer match. *J. Strength Cond. Res.* **2010**, *24*, 2683–2692. [CrossRef] [PubMed]
2. Stolen, T.; Chamari, K.; Castagna, C.; Wisloff, U. Physiology of soccer: An update. *Sports Med.* **2005**, *35*, 501–536. [CrossRef] [PubMed]
3. Barnes, C.; Archer, D.T.; Hogg, B.; Bush, M.; Bradley, P.S. The evolution of physical and technical performance parameters in the English Premier League. *Int. J. Sports Med.* **2014**, *35*, 1095–1100. [CrossRef] [PubMed]
4. Bush, M.; Barnes, C.; Archer, D.T.; Hogg, B.; Bradley, P.S. Evolution of match performance parameters for various playing positions in the English Premier League. *Hum. Mov. Sci.* **2015**, *39*, 1–11. [CrossRef] [PubMed]
5. Bradley, P.S.; Archer, D.T.; Hogg, B.; Schuth, G.; Bush, M.; Carling, C.; Barnes, C. Tier-specific evolution of match performance characteristics in the English Premier League: It's getting tougher at the top. *J. Sports Sci.* **2016**, *34*, 980–987. [CrossRef] [PubMed]

6. Nassis, G.P.; Massey, A.; Jacobsen, P.; Brito, J.; Randers, M.B.; Castagna, C.; Mohr, M.; Krustrup, P. Elite football of 2030 will not be the same as that of 2020: Preparing players, coaches, and support staff for the evolution. *Scand. J. Med. Sci. Sports* **2020**, *30*, 962–964. [CrossRef]
7. Carling, C.; Lacome, M.; McCall, A.; Dupont, G.; Le Gall, F.; Simpson, B.; Buchheit, M. Monitoring of Post-match Fatigue in Professional Soccer: Welcome to the Real World. *Sports Med.* **2018**, *48*, 2695–2702. [CrossRef]
8. Julian, R.; Page, R.M.; Harper, L.D. The Effect of Fixture Congestion on Performance during Professional Male Soccer Match-Play: A Systematic Critical Review with Meta-Analysis. *Sports Med.* **2021**, *51*, 255–273. [CrossRef]
9. Garcia, G.R.; Goncalves, L.G.C.; Clemente, F.M.; Nakamura, F.Y.; Nobari, H.; Bedo, B.L.S.; Azevedo, A.M.; Guerra, M.A., Jr.; Aquino, R. Effects of congested fixture and matches' participation on internal and external workload indices in professional soccer players. *Sci. Rep.* **2022**, *12*, 1864. [CrossRef]
10. Brito de Souza, D.; Lopez-Del Campo, R.; Resta, R.; Moreno-Perez, V.; Del Coso, J. Running Patterns in LaLiga Before and After Suspension of the Competition Due to COVID-19. *Front. Physiol.* **2021**, *12*, 666593. [CrossRef] [PubMed]
11. Noor, D.; McCall, A.; Jones, M.; Duncan, C.; Ehrmann, F.; Meyer, T.; Duffield, R. Perceived load, fatigue and recovery responses during congested and non-congested micro-cycles in international football tournaments. *J. Sci. Med. Sport* **2021**, *24*, 1278–1283. [CrossRef] [PubMed]
12. Dellal, A.; Lago-Penas, C.; Rey, E.; Chamari, K.; Orhant, E. The effects of a congested fixture period on physical performance, technical activity and injury rate during matches in a professional soccer team. *Br. J. Sports Med.* **2015**, *49*, 390–394. [CrossRef] [PubMed]
13. Dupont, G.; Nedelec, M.; McCall, A.; McCormack, D.; Berthoin, S.; Wisloff, U. Effect of 2 soccer matches in a week on physical performance and injury rate. *Am. J. Sports Med.* **2010**, *38*, 1752–1758. [CrossRef] [PubMed]
14. Lago-Penas, C.; Rey, E.; Lago-Ballesteros, J.; Casais, L.; Dominguez, E. The influence of a congested calendar on physical performance in elite soccer. *J. Strength Cond. Res.* **2011**, *25*, 2111–2117. [CrossRef] [PubMed]
15. Djaoui, L.; Wong, D.P.; Pialoux, V.; Hautier, C.; Da Silva, C.D.; Chamari, K.; Dellal, A. Physical Activity during a Prolonged Congested Period in a Top-Class European Football Team. *Asian J. Sports Med.* **2014**, *5*, 47–53. [CrossRef]
16. Doncaster, G.; White, P.; Svenson, R.; Page, R.M. The influence of fixture congestion on physical performance response to soccer match-play. *Res. Sports Med.* **2021**, *7*, 1–15. [CrossRef]
17. Djaoui, L.; Owen, A.; Newton, M.; Nikolaidis, P.T.; Dellal, A.; Chamari, K. Effects of congested match periods on acceleration and deceleration profiles in professional soccer. *Biol. Sport* **2022**, *39*, 307–317. [CrossRef]
18. Penedo-Jamardo, E.; Rey, E.; Padrón-Cabo, A.; Kalén, A. The impact of different recovery times between matches on physical and technical performance according to playing positions. *Int. J. Perform. Anal. Sport* **2017**, *17*, 271–282. [CrossRef]
19. Jones, R.N.; Greig, M.; Mawene, Y.; Barrow, J.; Page, R.M. The influence of short-term fixture congestion on position specific match running performance and external loading patterns in English professional soccer. *J. Sports Sci.* **2019**, *37*, 1338–1346. [CrossRef]
20. Lago-Ballesteros, J.; Lago-Penas, C.; Rey, E.; Casais, L.; Dellal, A. The effect of cumulative fatigue on activity profiles of professional soccer players during a congested fixture period. *Biol. Sport* **2010**, *27*, 181–185.
21. Zanetti, V.; Aoki, M.S.; Bradley, P.; Carling, C.; Marino, T.K.; Moreira, A. Running Performance and Hormonal, Maturity and Physical Variables in Starting and Non-Starting Elite U14 Soccer Players During a Congested Match Schedule. *J. Hum. Kinet.* **2021**, *80*, 287–295. [CrossRef] [PubMed]
22. Zanetti, V.; Carling, C.; Aoki, M.S.; Bradley, P.S.; Moreira, A. Are There Differences in Elite Youth Soccer Player Work Rate Profiles in Congested vs. Regular Match Schedules? *J. Strength Cond. Res.* **2021**, *35*, 473–480. [CrossRef] [PubMed]
23. Castellano, J.; Martin-Garcia, A.; Casamichana, D. Most running demand passages of match play in youth soccer congestion period. *Biol. Sport* **2020**, *37*, 367–373. [CrossRef] [PubMed]
24. Jimenez, S.L.; Mateus, N.; Weldon, A.; Bustamante-Sanchez, A.; Kelly, A.L.; Sampaio, J. Analysis of the most demanding passages of play in elite youth soccer: A comparison between congested and non-congested fixture schedules. *Sci. Med. Footb.* **2022**, *29*, 1–8. [CrossRef]
25. Andrzejewski, M.; Konarski, M.J.; Chmura, J.; Pluta, B. Changes in the activity profiles of soccer players over a three-match training micro cycle. *Int. J. Perform. Anal. Sport* **2014**, *14*, 814–828. [CrossRef]
26. Varley, M.C.; Di Salvo, V.; Modonutti, M.; Gregson, W.; Mendez-Villanueva, A. The influence of successive matches on match-running performance during an under-23 international soccer tournament: The necessity of individual analysis. *J. Sports Sci.* **2018**, *36*, 585–591. [CrossRef]
27. Rampinini, E.; Impellizzeri, F.M.; Castagna, C.; Abt, G.; Chamari, K.; Sassi, A.; Marcora, S.M. Factors influencing physiological responses to small-sided soccer games. *J. Sports Sci.* **2007**, *25*, 659–666. [CrossRef]
28. Di Salvo, V.; Pigozzi, F.; Gonzalez-Haro, C.; Laughlin, M.S.; De Witt, J.K. Match performance comparison in top English soccer leagues. *Int. J. Sports Med.* **2013**, *34*, 526–532. [CrossRef]
29. Al Haddad, H.; Mendez-Villanueva, A.; Torreno, N.; Munguia-Izquierdo, D.; Suarez-Arrones, L. Variability of GPS-derived running performance during official matches in elite professional soccer players. *J. Sports Med. Phys. Fit.* **2018**, *58*, 1439–1445. [CrossRef]
30. Doncaster, G.; Unnithan, V. Between-Game Variation of Physical Soccer Performance Measures in Highly Trained Youth Soccer Players. *J. Strength Cond. Res.* **2019**, *33*, 1912–1920. [CrossRef]

31. Oliva-Lozano, J.M.; Muyor, J.M.; Fortes, V.; McLaren, S.J. Decomposing the variability of match physical performance in professional soccer: Implications for monitoring individuals. *Eur. J. Sport Sci.* **2021**, *21*, 1588–1596. [CrossRef] [PubMed]
32. Arruda, A.F.; Carling, C.; Zanetti, V.; Aoki, M.S.; Coutts, A.J.; Moreira, A. Effects of a very congested match schedule on body-load impacts, accelerations, and running measures in youth soccer players. *Int. J. Sports Physiol. Perform.* **2015**, *10*, 248–252. [CrossRef] [PubMed]
33. Mortatti, A.L.; Oliveira, R.S.C.; Pinto, J.; Galvao-Coelho, N.L.; de Almeida, R.N.; Aoki, M.S.; Moreira, A. A Congested Match Schedule Alters Internal Match Load and Affects Salivary Immunoglobulin A Concentration in Youth Soccer Players. *J. Strength Cond. Res.* **2022**, *36*, 1655–1659. [CrossRef]
34. Pinto, J.; de Oliveira, R.S.C.; Galvao-Coelho, N.L.; de Almeida, R.N.; Moreira, A.; Mortatti, A.L. The Effects of Successive Soccer Matches on the Internal Match Load, Stress Tolerance, Salivary Cortisol and Jumping Performance in Youth Soccer Players. *J. Hum. Kinet.* **2021**, *80*, 173–184. [CrossRef] [PubMed]
35. Mohr, M.; Krustrup, P.; Bangsbo, J. Match performance of high-standard soccer players with special reference to development of fatigue. *J. Sports Sci.* **2003**, *21*, 519–528. [CrossRef] [PubMed]
36. Di Salvo, V.; Baron, R.; Tschan, H.; Calderon Montero, F.J.; Bachl, N.; Pigozzi, F. Performance characteristics according to playing position in elite soccer. *Int. J. Sports Med.* **2007**, *28*, 222–227. [CrossRef] [PubMed]
37. Soroka, A.; Lago-Peñas, C. The effect of a succession of matches on the physical performance of elite football players during the World Cup Brazil 2014. *Int. J. Perform. Anal. Sport* **2016**, *16*, 434–441. [CrossRef]
38. Williams, A.M.; Ford, P.R.; Drust, B. Talent identification and development in soccer since the millennium. *J. Sports Sci.* **2020**, *38*, 1199–1210. [CrossRef]
39. Dodd, K.D.; Newans, T.J. Talent identification for soccer: Physiological aspects. *J. Sci. Med. Sport* **2018**, *21*, 1073–1078. [CrossRef]
40. Winter, E.M.; Maughan, R.J. Requirements for ethics approvals. *J. Sports Sci.* **2009**, *27*, 985. [CrossRef]
41. Sausaman, R.W.; Sams, M.L.; Mizuguchi, S.; DeWeese, B.H.; Stone, M.H. The Physical Demands of NCAA Division I Women's College Soccer. *J. Funct. Morphol. Kinesiol.* **2019**, *4*, 73. [CrossRef] [PubMed]
42. Asian-Clemente, J.; Rabano-Munoz, A.; Requena, B.; Suarez-Arrones, L. High-speed training in a specific context in soccer: Transition games. *Int. J. Sports Med.* **2022**, *43*, 881–888. [CrossRef] [PubMed]
43. Asian-Clemente, J.; Rabano-Munoz, A.; Munoz, B.; Franco, J.; Suarez-Arrones, L. Can Small-side Games Provide Adequate High-speed Training in Professional Soccer? *Int. J. Sports Med.* **2021**, *42*, 523–528. [CrossRef]
44. Hopkins, W.G.; Marshall, S.W.; Batterham, A.M.; Hanin, J. Progressive statistics for studies in sports medicine and exercise science. *Med. Sci. Sports Exerc.* **2009**, *41*, 3–13. [CrossRef]
45. Ferraz, R.; Goncalves, B.; Coutinho, D.; Marinho, D.A.; Sampaio, J.; Marques, M.C. Pacing behaviour of players in team sports: Influence of match status manipulation and task duration knowledge. *PLoS ONE* **2018**, *13*, e0192399. [CrossRef] [PubMed]
46. Johnston, R.J.; Watsford, M.L.; Austin, D.; Pine, M.J.; Spurrs, R.W. Player acceleration and deceleration profiles in professional Australian football. *J. Sports Med. Phys. Fit.* **2015**, *55*, 931–939.
47. Silva, H.; Nakamura, F.Y.; Beato, M.; Marcelino, R. Acceleration and deceleration demands during training sessions in football: A systematic review. *Sci. Med. Footb.* **2022**, *26*, 1–16. [CrossRef] [PubMed]
48. Martin-Garcia, A.; Casamichana, D.; Diaz, A.G.; Cos, F.; Gabbett, T.J. Positional Differences in the Most Demanding Passages of Play in Football Competition. *J. Sports Sci. Med.* **2018**, *17*, 563–570.
49. Riboli, A.; Semeria, M.; Coratella, G.; Esposito, F. Effect of formation, ball in play and ball possession on peak demands in elite soccer. *Biol. Sport* **2021**, *38*, 195–205. [CrossRef]
50. Jaspers, A.; Kuyvenhoven, J.P.; Staes, F.; Frencken, W.G.P.; Helsen, W.F.; Brink, M.S. Examination of the external and internal load indicators' association with overuse injuries in professional soccer players. *J. Sci. Med. Sport* **2018**, *21*, 579–585. [CrossRef]
51. Aquino, R.; Carling, C.; Palucci Vieira, L.H.; Martins, G.; Jabor, G.; Machado, J.; Santiago, P.; Garganta, J.; Puggina, E. Influence of Situational Variables, Team Formation, and Playing Position on Match Running Performance and Social Network Analysis in Brazilian Professional Soccer Players. *J. Strength Cond. Res.* **2020**, *34*, 808–817. [CrossRef] [PubMed]
52. Augusto, D.; Brito, J.; Aquino, R.; Figueiredo, P.; Eiras, F.; Tannure, M.; Veiga, B.; Vasconcellos, F. Contextual Variables Affect Running Performance in Professional Soccer Players: A Brief Report. *Front. Sports Act. Living* **2021**, *3*, 778813. [CrossRef] [PubMed]
53. Gonçalves, L.G.C.; Clemente, F.; Palucci Vieira, L.H.; Bedo, B.; Puggina, E.F.; Moura, F.; Mesquita, F.; Pereira Santiago, P.R.; Almeida, R.; Aquino, R. Effects of match location, quality of opposition, match outcome, and playing position on load parameters and players' prominence during official matches in professional soccer players. *Hum. Mov.* **2021**, *22*, 35–44. [CrossRef]
54. Palucci Vieira, L.H.; Carling, C.; Barbieri, F.A.; Aquino, R.; Santiago, P.R.P. Match Running Performance in Young Soccer Players: A Systematic Review. *Sports Med.* **2019**, *49*, 289–318. [CrossRef]
55. Arjol-Serrano, J.L.; Lampre, M.; Diez, A.; Castillo, D.; Sanz-Lopez, F.; Lozano, D. The Influence of Playing Formation on Physical Demands and Technical-Tactical Actions According to Playing Positions in an Elite Soccer Team. *Int. J. Environ. Res. Public Health* **2021**, *18*, 41–48. [CrossRef]
56. Forcher, L.; Hartel, S.; Jekauc, D.; Wasche, H.; Woll, A.; Gross, T.; Altmann, S. Does Technical Match Performance in Professional Soccer Depend on the Positional Role or the Individuality of the Player? *Front. Psychol.* **2022**, *13*, 813206. [CrossRef]
57. Forcher, L.; Jekauc, D.; Woll, A.; Gross, T.; Altmann, S. Center backs work hardest when playing in a back three: The influence of tactical formation on physical and technical match performance in professional soccer. *PLoS ONE* **2022**, *17*, e0265501. [CrossRef]

58. Bulgay, C.; Cetin, E.; Orhan, O.; Ergun, M.A. The effects of the ACTN3 and ACE genes on the sportive performance of athletes. *Inonu Univ. J. Phys. Educ. Sport Sci.* **2020**, *7*, 1–12.
59. Altmann, S.; Neumann, R.; Woll, A.; Hartel, S. Endurance Capacities in Professional Soccer Players: Are Performance Profiles Position Specific? *Front. Sports Act. Living* **2020**, *2*, 549897. [CrossRef]

MDPI
St. Alban-Anlage 66
4052 Basel
Switzerland
www.mdpi.com

Journal of Functional Morphology and Kinesiology Editorial Office
E-mail: jfmk@mdpi.com
www.mdpi.com/journal/jfmk

www.ingramcontent.com/pod-product-compliance
Lightning Source LLC
Chambersburg PA
CBHW061607210926
51850CB00009B/2236
9783036587790